Business Process Management

A Dramatized Journey

Sandeep Purao
Bentley University

Caroline Strange

Prospect Press

Founded in 2014, Prospect Press serves the academic discipline of Information Systems by publishing essential and innovative textbooks across the curriculum including introductory, emerging, and upper-level courses. Based in Burlington, Vermont, Prospect Press distributes titles worldwide. We welcome new authors to send proposals or inquiries to Beth.golub@ProspectPressVT.com.

Editor: Beth Lang Golub
Associate Editor: Dave Williams
Production Management: Scribe Inc.
Cover Design and Illustrations: Annie Clark
Conference table image on cover based on iStock.com/471514178
Author photographs Copyright Sandeep Purao and Caroline Strange

eTextbook:
Edition 1.0
ISBN: 978-1-958303-14-6
Available from Redshelf, VitalSource, and Perusall

Printed Paperback:
Edition 1.0
ISBN: 978-1-958303-15-3
Available from Redshelf

For more information, visit
https://www.prospectpressvt.com/textbooks/purao-business-process-management

Contents

PART 1. THE SETUP

PART 2. THE PLAY

Act I. Exploring Processes in the Organization

Act II. Working with a Process

Act III. Considering Technological Solutions

*Couplets inspired by the work of Vikram Seth, A Suitable Boy

Preface

In an era when business landscapes are rapidly transforming, understanding the intricacies of business process management (BPM) becomes imperative for anyone aspiring to lead or innovate in this space. *Business Process Management: A Dramatized Journey* is an unconventional book that uses the power of drama and narrative to unravel the complexities of BPM.

This book stands at the crossroads of academic rigor and creative storytelling, presenting BPM not as a static set of principles but as a dynamic, human-centric journey. Through a series of ten scenes, each mirroring a stage in the BPM life cycle, we aim to make the learning process not only informative but also engaging and relatable.

The Approach

Our approach is simple yet impactful. We believe that the essence of learning lies in engagement, and what better way to engage than through stories and characters that readers can connect with? Each scene in this book is a window into a stage of BPM, crafted to bring theoretical concepts to life through the experiences, struggles, and triumphs of our characters.

Target Audience

The book is designed for BPM courses at either the undergraduate or graduate level within a business or information systems (IS) curriculum. The only prerequisite for the book is some understanding of what business organizations do and how they are structured. The book can be used in several ways:

- As part of a one-semester course focused on BPM within the IS curriculum
- As part of a one-semester course on BPM with a more technical focus to introduce an organizational perspective
- As part of a one-semester course on digital transformation to include process improvements as a path to digital transformation
- As part of a one-semester course on advanced BPM to emphasize the organizational perspective

The book can also be used by business managers interested in introducing BPM initiatives in their organization—as well as process management professionals interested in understanding how to work with organizations to achieve process improvement goals.

The Structure

Each of the ten scenes is short, making them ideal for classroom settings or independent study. They cover the gamut of BPM stages—from the initial steps of process orientation and process identification to the more intricate aspects of process design, analysis, and implementation. Each scene takes place in a fictional company, chosen as a relatable backdrop against which the drama of BPM unfolds.

Characters: The Heart of Our Story

At the heart of each scene are our characters—a diverse mix of individuals from various echelons within the company, alongside external vendor representatives. These characters are not mere fictional constructs; they are carefully designed to represent the different stakeholders typically involved in BPM initiatives. By following their journeys, readers are exposed to a multitude of perspectives, each bringing its own set of challenges and insights into the BPM process.

Interactive and Reflective Learning

The interactive nature of the scenes is further enhanced by the inclusion of reflection questions and self-assessment exercises at the end of each. These are not just afterthoughts; they are integral components that encourage students to engage critically with the material. They serve as catalysts for discussion, deepening understanding and fostering a culture of inquiry and introspection.

Supplementary Resources

To ensure that the dramatic narrative is grounded in practical reality, some scenes are supplemented with detailed examples (which appear as tables or appendixes). At the end of each scene, we also provide a list of key terms. These resources serve to bridge the gap between theory and practice, providing a solid foundation to support reflection and consolidate learning from the dramatized stories.

Educational Use

While designed to be versatile enough for various educational settings, this book is particularly effective when used alongside traditional textbooks or readings. It's a supplement, not a substitute. The scenes are best experienced when read aloud in groups, allowing students to actively engage with the material and one another. This method of collaborative learning is not just about understanding BPM; it's about experiencing it. During the two-plus semesters that we have tested the book in the classes we have taught, we have learned that there are many ways the students engage with the book by taking on the roles of the characters and sometimes agreeing with them, sometimes not.

As an educator, practitioner, or student, this book invites you to embark on a journey unlike any other—a journey where learning is not confined to the pages of a textbook but brought to life through the power of drama and storytelling. *Business*

Process Management: A Dramatized Journey is more than just a book; it's a new way to experience and understand the world of BPM.

Welcome aboard this unique journey. Let each page turn be a step forward in your experience of BPM. Let the play begin!

Additional Instructor Resources

The following materials are included in the instructor resources, which are available at http://prospectpressvt.com/textbooks/purao-business-process-management:

- PowerPoint presentations: Instructors can access short PowerPoint presentations (to introduce the book and each scene) that they can use as is or customize.
- Instructions for reading the scripts aloud: Dramatized scripts can have the best impact when they can be read aloud. Instructors can access and share instructions for how to do this.
- Suggested answers to reflection questions: Instructors can access suggested answers to reflection questions with the acknowledgment that these represent one potentially appropriate answer.

We hope you enjoy reading and using the book as much as we have enjoyed writing it.

Acknowledgments

Writing a book is never an easy task. It is even more demanding when it requires mixing two dramatically (pun intended) different traditions—academic rigor and creative storytelling—and packaging the outcome as dramatized scripts. That we decided to take this on as first-time book authors was one of the many unusual choices that we all made during the COVID-19 pandemic. In retrospect, it looks like we did make some good use of the time.

We want to acknowledge the significant influence of students in undergraduate and graduate courses about BPM at Bentley University across multiple years. Discussions with many of them have shaped the examples, the characters, and the fictitious company we have devised in the book. Over the last two semesters, as we continued to develop and refine the dramatized scripts, we used individual scripts and sometimes a subset of these across multiple semesters. This allowed us to identify any missing details and correct inconsistencies. The discussions sometimes also surfaced new ideas that we incorporated in the dramatized scripts. We would also like to acknowledge the influence of discussions with colleagues at Bentley University over several years, which has allowed us to structure the book.

Finally, we would like to thank the team at Prospect Press, and particularly, Beth Lang Golub, for accepting our somewhat outlandish ideas about writing a play to teach BPM. Thank you, Beth, Dave Williams, and the entire Prospect Press team for allowing us to participate in so many of the steps in the process and for handling our questions and interactions throughout the journey.

Testimonials

Purao and Strange's text on business processing management (BPM) is a departure from the norm for teaching students about this critical topic. Instead of having students just read about how BPM works, students *enact* their learning through a drama. The dramatic scenes move one through the typical stages of introducing BPM in a real-world company.

The term-long play begins, appropriately enough, with introductions of the "actors" (or "dramatis personae"). These range from vice presidents of different divisions and functions to professionals in the focal company, Royal Energy. Variations in personalities reflect the diversity of people who would typically play out roles in a reengineering effort. As the scenes develop, participants hone in on important distinctions, such as how processes differ from functions and departments. As the knowledge base grows, the play illuminates how processes can be modeled (intellectually and technically) and how genuine cooperation can lead to more efficient and understandable processes.

What students gain from this experience is learning through enactment, or perhaps more precisely stated, simulated participation in real-world scenarios. Through the immediacy of role-playing, they see how processes can be better modeled and how a redesign of an organization's operations can lead to immense organizational value. This book represents a vitally new pedagogy for conveying intricate material to students, and I urge faculty to consider the value of this form of *enactment* of learning in their classes.

—Detmar Straub
Professor, Temple University
November 6, 2023

This book takes students on a novel and fun journey through the human side of business process management and technology implementation. It helps build student intuition from the bottom up by exposing them to the messy social situations through which work is transformed in real-life settings. Along the way, students develop familiarity with techniques for building and analyzing process models and tackling organizational problems for deploying processes in organizations.

—M. Lynne Markus
Bentley University

This book is a great addition to BPM education. I love the way it is written as a play and how creatively it can be used. Not only is this new format refreshingly different and engaging, it also fulfills a most valuable didactic purpose: as BPM is not a routine task, textbook knowledge falls short on providing the important, comprehensive BPM skill set. BPM requires skills to navigate complexity, to creatively find innovative solutions to real-world problems, and to communicate and discuss ideas and results. To this end, the book offers a whole new realm of experience: students deal with a complex context in a playful (and narrative) way and are guided to understand these from the process perspective and—above all—to develop and discuss creative solutions to problems. I have already had the privilege of trying out this book in an early state, and it was amazing to see how it works. I will for sure continue to use it as an important component of my BPM education portfolio in the future. Congratulations to Sandeep, who has brilliantly combined his passion for both process and design science in this piece.

—Jan vom Brocke
Professor at Muenster and Director
of European Research Center for IS

This is certainly not just another BPM book—this unique and engaging screenplay-style introduction to business process management is the perfect companion for flipping your classroom and making BPM learning fun and interactive. Follow the main characters as they grapple with identifying, prioritizing, and improving processes; assign your students to characters to enable them to appreciate different perspectives; or simply use the scenes in the play together with the included reflection questions to facilitate active class discussion. Say goodbye to asking your BPM class a question only to be met with silence in response!

—Marta Indulska
Professor, University of Queensland

About the Authors

Sandeep Purao is on faculty in the Business School at Bentley University, having previously worked at Penn State University and Georgia State University. His research involves the design of IT-based solutions for societal problems. Outcomes from his work have been published in scholarly journals and conferences and have been supported by public and private funding agencies. He holds a PhD in management science from the University of Wisconsin-Milwaukee.

Caroline Strange is an Alaskan-born actress and singer working in New York City. She has worked across several commissioned works both online and onstage, including off-Broadway productions and repertory theaters in different states as well as commercials for diverse audiences. Caroline enjoys doing voice work, script writing, and plays in different genres. She holds a master of fine arts in acting from the University of North Carolina at Chapel Hill.

PART 1
The Setup

Prologue

Here are some good folks on a BPM odyssey
Let's join them—what are they up to, let's see!

The Organization

Royal Energy is a long-standing company in New England. In service since the early 1960s, their main business is serving energy to about four million residential and commercial customers. The company has four divisions: Eastern MA, Western MA, New Hampshire, and Rhode Island. Regina Wood is the vice president for the Eastern MA division, which includes Greater Boston. Samantha Bellman runs Western MA, including Worcester. The New Hampshire division, which includes Portsmouth, is led by Anik Malik. And David Ashley oversees the Rhode Island division, which includes the city of Providence. Each division manages its own functions, such as HR, Finance, Operations, and Marketing. There is standardization across divisions because of the regulated nature of the industry. State mandates make the operations in New Hampshire and Rhode Island different from the two divisions in Massachusetts. Over the six decades, the structures and operations at Royal Energy have been stable. The following chart shows the structure of the organization and the key characters who are about to play a role in this journey.

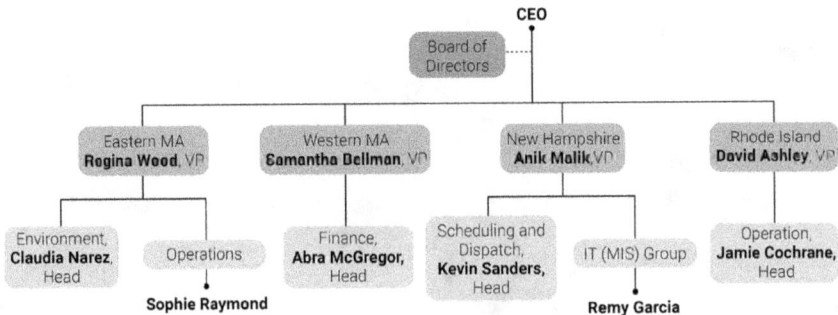

CEO

Board of Directors

Eastern MA	Western MA	New Hampshire	Rhode Island
Regina Wood, VP	**Samantha Bellman**, VP	**Anik Malik**, VP	**David Ashley**, VP

Environment, **Claudia Narez**, Head — Operations — **Sophie Raymond**

Finance, **Abra McGregor**, Head

Scheduling and Dispatch, **Kevin Sanders**, Head — IT (MIS) Group — **Remy Garcia**

Operation, **Jamie Cochrane**, Head

The Industry

The energy industry is at the heart of modern economies. Many of us in industrialized economies consume large amounts of energy or are served by businesses and cultural organizations that rely on the uninterrupted flow of energy. The extraction and supply of energy are, therefore, crucial to work and leisure. The industry focuses on the generation and distribution of energy to residential and commercial customers. Locational advantages and the need for exploration have meant that most firms in the industry specialize in different parts of the supply chain, such as extraction, manufacturing and refining, and distribution. Firms that claim membership in the industry work with fossil fuels (petroleum and oil, refiners, fuel transport), coal (extraction and processing), natural gas (extraction and distribution), electrical power (generation and distribution), nuclear power (plants and distribution), and renewables (such as hydroelectric, wind, and solar). As the industry has been criticized for causing pollution, global warming, and other environmental problems, it is turning away from fossil fuels to invest in renewable and sustainable energy. With significant growth across the planet, including emerging economies such as China and India, the demand for energy continues unabated. The industry continues to change, with a mix of residential and commercial customers as well as varied sources of energy, including coal, water, wind, and solar. It is against this backdrop that we see Royal Energy engage with the problems of appreciating, analyzing, redesigning, implementing, and rolling out business processes.

The Cast of Characters

DIVISION HEADS

Regina Wood, VP, Eastern MA

Regina heads the Eastern MA division. Her colleagues see her as calm and trustworthy. She is a woman of few words and not prone to drama. When she speaks, people listen. While integrity and logic rule in her book, she still gives special attention to the opinions and personalities of her team. Regina holds a degree in communications from Howard and has consistently made vertical career moves. She lives in Arlington, MA, with her husband and three kids.

(She/Her/Hers)

Samantha Bellman, VP, Western MA

Samantha, fashionable and a little quirky, runs the Western MA division. She has an unapologetic personality and is valued for her boldness and intellect. After graduating from Williams College, she built a career at Royal Energy. Her home is Worcester, MA. Her kids are grown up, and she lives with her wife in a quiet suburb.

(She/Her/Hers)

Anik Malik, VP, New Hampshire

Often the instigator, Anik seems to thrive on challenging the ideas of his colleagues. Nevertheless, his peers and teammates respect him and consider him a worthy opponent in an argument. After getting dual degrees in business and engineering from Wisconsin, Anik has had a successful career in the company. He lives in Manchester, NH, with his husband and teen-aged daughter.

(He/Him/His)

David Ashley, VP, Rhode Island

Born and raised in Georgia, David has relied on his southern charisma to make friends throughout the company, somehow landing in a position that lets his talents really shine. He has been with Royal Energy for a few years now, having worked in Texas after getting a degree in English from Georgia Tech. David is a lifelong bachelor and maintains a condo near Newport, RI.

(He/Him/His)

DEPARTMENTAL REPRESENTATIVES
Claudia Narez, Environmental, Eastern MA

Manager of the Environmental group for Eastern MA, Claudia is a natural leader. After getting her dual degree in business management and environmental law from Monterrey, Mexico, she moved north to work with Royal Energy and quickly earned the loyalty and respect of her staff. However, in spite of her consistency and clear boundaries, Claudia feels that she does not get the recognition she deserves for running a tight ship. She lives in Cambridge with her husband and a toddler.

(She/Her/Hers)

Jamie Cochrine, Operations, Rhode Island

Jamie runs the Operations group in Rhode Island with a hands-on intellect that has proven successful in his work within the company. However, he is still learning mature management skills. As a new manager, he often walks on eggshells with his staff and has not yet cultivated trust with them. He has been in his position only a few months and is wary of introducing changes. Jamie is a graduate of Brown University, with a self-designed bachelor's degree in sustainable business. He lives in Providence with his partner and is known to be active in his local board game community.

(He/Him/His)

Kevin Sanders, Scheduling/Dispatch, New Hampshire

Kevin's long tenure at Royal Energy may be due to his jovial, curious attitude and willingness to advise his younger colleagues. However, he sometimes lacks the tact and follow-through to be helpful in a tense situation. Kevin's college work was done at Suffolk in Boston. After that, he moved back to New Hampshire and joined Royal Energy. After a few lateral moves, he now runs Marketing for New Hampshire. Kevin lives in Manchester and has been married to his high school sweetheart for thirty-two years.

(He/Him/His)

Abra McGregor, Finance, Western MA

Abra McGregor runs Finance for Western MA. She is highly organized, runs her staff like a well-oiled machine, and is known to work long hours. She is a private, unemotional team member who is often seen taking lots of notes in meetings rather than engaging. She can be resistant to change (she opposes new payment forms like Bitcoin) but is quite capable of rolling out new initiatives once convinced. Her college years were spent in North Carolina, where she studied business and theater. She lives in a small community north of Worcester, MA, with her pets.

(She/Her/Hers)

OPERATIONAL AND IT EXPERTS

Sophie Raymond, Operations, Eastern MA

Growing up in Kamloops, BC, and raised by her grandfather, Sophie did not think she would be working in Boston one day. She attended Bentley University for her degree in data analytics and fell in love with the breadth of experiences. Soon after she graduated, Sophie's grandfather passed away. She chose MA to be her permanent home due to her involvement with the local outdoor adventure community. On the weekends, she is seen hiking or skiing, depending on the season. She lives in Watertown with her elderly cat and many, many plants.

(She/Her/Hers)

Remy Garcia, IT, New Hampshire

A Spaniard who arrived in Boston for his degree in computer science from Tufts, Remy was another who decided to stay on and make Boston home. However, he has retained a certain European flair. He is sharply dressed and combines his strong technology skills with natural empathy and good humor to achieve his goals. On a personal level, he often talks about how much he misses Barcelona's good tapas and nightlife. He lives alone in Portsmouth at one of those new, high-tech condos.

(He/Him/His)

VENDOR REPRESENTATIVES

Matthias Schuster, TBQ, Enterprise Systems Vendor

Matthias has been with TBQ for a while now. He spent his early career years in Frankfurt, Germany, after getting his degree in communications from Heidelberg University. In their thirties, Matthias moved back to the US and now roams the Eastern Seaboard as a reliable and knowledgeable representative for TBQ. Family oriented, Matthias bought a home near his parents in Charlotte, NC, which allows him to travel up and down the coast as needed.

(He/They)

Shalini Sharma, Sonita, BPMS Vendor

Shalini represents Sonita, a strong BPMS vendor. She grew up in Birmingham, UK, with East Indian parents and earned a scholarship to the University of Warwick near Coventry. She is now the lead representative for Sonita in the Northeast US. Life in New York is different from Birmingham, but she is thriving in this role, developing a reputation not only with her own company but also in the professional community. Shalini's friends considered her a "bossy know-it-all" growing up. Now colleagues across organizations see her as becoming one of the thought leaders in the industry, and she lives up to those expectations.

(She/Her/Hers)

The Timeline

The scenes in this book are designed to provide you with glimpses of the journey for Royal Energy. In the timeline below, each scene is shown at the bottom with an indication of time elapsed (not to scale). The middle layer shows you the participants in each scene. At the top, the callouts point to the outcomes of work done by different teams during the time between scenes.

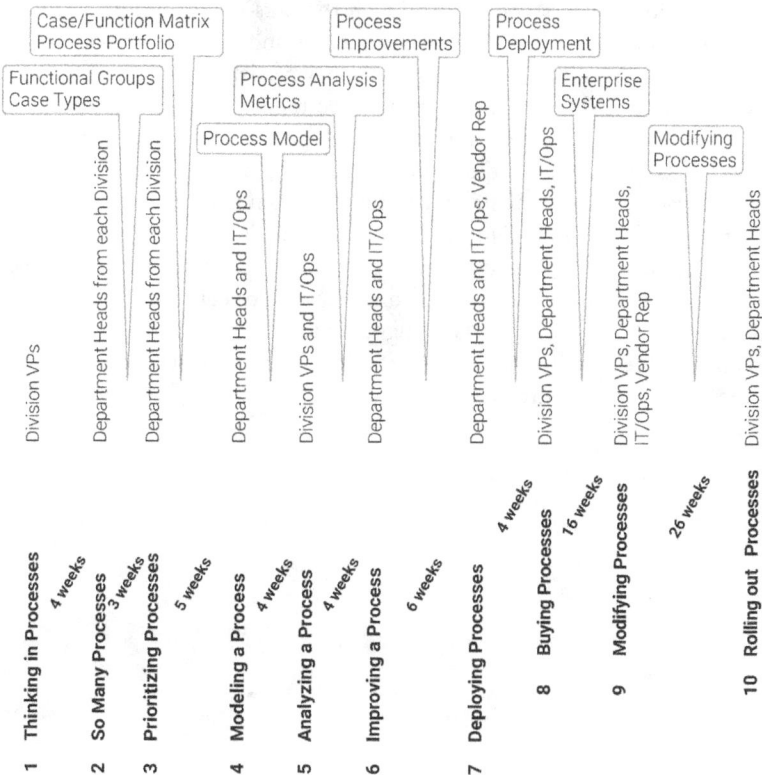

Timeline (about eighteen months)

How to Use This Book

The play, organized as ten scenes across three acts, provides glimpses of the journey for Royal Energy. Although fictional, the company and the roles are inspired by experiences and discussions with students and colleagues. They are intended to add authenticity without being overwhelming.

The scenes allow an exploratory pedagogy. Each provides the students with a focused set of concerns. By taking on the role of any of the characters in a scene, the students attempt to understand different perspectives and evaluate their own understanding. Some scenes point to multiple possibilities without offering a clear consensus, allowing the instructor to lead the conversation. A set of reflection questions at the end of each scene allows the students greater engagement with situated concerns as well as the larger concepts. The scenes may be used in several ways in a graduate / advanced undergraduate class about business process management.

Plan 1. Assign a scene for the students to read before the class session. A part of the class session can then be structured as a discussion using the reflection questions. In each class, the instructor may select scenes to allow the students to appreciate different sets of concerns. Although the scenes are arranged as a sequence in time, it is possible to assign the scenes out of sequence because each includes an introductory narration that provides some context.

Plan 2. Assign a scene for the students to read before the class session. During the class session, invite students to take on the personas of different characters. This would allow them to consider the concerns and opportunities as seen from different perspectives. The roles for students may persist or vary, allowing a student to participate with different perspectives. The reflection questions may then be considered from the vantage point of different characters.

Plan 3. Assign a scene for the students to read before the class session. During the class session, invite students to describe how such a discussion may take place in their organization. This plan can be useful particularly when the students have some experience so they can contribute based on a specific personality they may have encountered. The reflection questions may then be considered as an initial set, with an invitation to the students to suggest additional concerns.

You may come up with more creative ways to use the scenes, a group of scenes as an act, or even the entire play.

Enjoy!

PART 2
The Play

ACT I

Exploring Processes in the Organization

Scene 1
Thinking in Processes

Facing a challenge, the division leaders meet one day
As the discussion proceeds, process thinking holds sway

Learning Objectives

- Describe what a business process is and is not
- Identify the difference between a business process and a business function
- Explain why most organizations are structured around business functions
- Argue about the importance of focusing on business processes

Characters

Regina Wood—VP, Eastern MA
Anik Malik—VP, New Hampshire
Samantha Bellman—VP, Western MA

Narrator: *Three of the four VPs of Royal Energy are meeting at the Hyatt Regency in Cambridge to discuss a problem. David Ashley, the Rhode Island VP, is unable to attend. But this meeting cannot be pushed back. The dining room on the club level is fairly quiet, with only a couple of people checking emails as they enjoy breakfast. Regina is already sitting at the table, grateful to be in a familiar setting but anticipating an uncomfortable morning. Regina knows that sometimes what goes on behind the scenes is just as important as the work and operations. Knowing that Samantha and Anik do not always see eye to eye, she hopes that today will not* *be a repeat of their last meeting. Samantha arrives and greets Regina warmly.*

SAMANTHA: Good morning, Regina! It's good to see you again. How are things?

REGINA: Hello, Samantha. Good to see you too. I took the liberty of putting in an order of croissants and coffee for the table; the waiter should be back soon.

SAMANTHA: Ooh, the chocolate ones? Or almond?

REGINA: Both.

SAMANTHA: Perfect!

Narrator: *Samantha makes herself comfortable at the table. Regina is looking forward to hearing her perspective on the reason they're meeting and is about to tell her so when Anik arrives.*

ANIK: Good morning, Regina. Samantha. Oh . . . the Hyatt's impression of what a croissant is. These must be here for your benefit, right, Sam?

Narrator: *Samantha and Regina share a brief glance as they all settle in to order breakfast. Samantha may be the usual recipient of Anik's goading, but she is never the victim. Initially tense, the table begins to warm up. Regina talks about her mom's recent knee surgery, and Anik laughs about how he and his wife are repainting their newly renovated bathroom for the second time. According to their teenager, forest green is bad for reflecting light during makeup application, and they have found she is right. We return to the group as Samantha talks about finding a new social media manager for her division.*

SAMANTHA: Oh, it's been impossible. Ours was so good, she left us to start her own company managing content for an RPG online store.

REGINA: RPG?

ANIK: My daughter says it stands for Really Pretentious Gamers, but I think that's not right . . .

SAMANTHA: *Anyway* . . . turns out replacing her isn't easy, and I can see why she left! I have customers complaining about service interruptions because they can't reach a human representative using the website help center, which is still a mess, by the way! We're getting a new person tomorrow, fingers crossed—but honestly? A new social media manager is turning out to be the least of my worries.

REGINA: Well, this is what we're here to talk about, isn't it? I'm having similar issues in my region. My staff is constantly overwhelmed, and frankly, so am I. We have these legacy-era reports that take up so much time. I mean, I get that the reports are important, but this is getting out of hand.

ANIK: Right, internal reporting, regulatory reports, reports just because . . .

REGINA: Exactly, and it seems to be an endless stream.

Narrator: *The group continues to share the issues they're facing. Regina bemoans the struggle of only being in her position for a year and spending much of it writing reports instead of responding to customers. Samantha notes that the different departments in her division can't even ask one another for help because their reporting requirements are different.*

SAMANTHA: I wish we could do something like . . . streamline things across divisions, you know? Is that even a thing?

REGINA: I'm not sure, but sometimes it feels like we're basically just keeping the lights on, and barely.

ANIK: Oh, c'mon, it's not *that* bad . . .

REGINA: But you know what I mean, right? It was okay in the past—if we dial it back to the time when the company was smaller, maybe. But at the rate we're growing, this is not sustainable! We'll all just become glorified report writers.

Narrator: *Anik adds that there are people in his department with great ideas for new products and service offerings, but they've had no time to cultivate this creativity. Samantha agrees, saying . . .*

SAMANTHA: My people have been coming up with good ideas too, and there are so many that we'd love to start talking about. Like the progress with new energy sources, smart grids . . .

ANIK: Consumer-driven renewables are really popular right now. And where are we in that space? Nowhere!

REGINA: And we need to be on top of these ideas, but our functional heads are all only focused on their piece of the puzzle. Consider the individuals in Marketing, Finance, or, uh . . . what's your friend's name in my Sales Department?

SAMANTHA: Magdelyn.

REGINA: That's right. All so talented, but they can't even look past their own departments. Ultimately, we're ignoring the customer. We need to find a way to get all this talent to work together.

ANIK: Regina, I don't think that's the problem. Honestly, we're just short-staffed. I've been saying it for years! We just need to hire more people to take care of those reports. Or, better yet, why don't we get some specialists who will be responsible for customer service?

SAMANTHA: Well, I completely disagree with that. It's not how many people are on a task or how specialized they are. In fact, that itself may be the problem! We can't get our functional heads to work together. Why hire more people when the issue isn't to saturate our workforce but to—

ANIK: No, you clearly don't understand—

SAMANTHA: I wasn't finished. (*she waits for a moment*) Thank you. And yes, actually, I do understand. We all know this. A simple "fix" is to hire more workers and specialists. Frankly, it's how this company has always resolved issues in the past. And it might have helped in a market when we knew our customer base, we knew exactly what we were selling, and it was a matter of doing things faster. Now we have new products we don't know how to develop, customers who want to behave like suppliers, and technologies that make it all possible—what do we make of this? We're not going to be able to do the same old thing in the long term! And why should we be resistant to trying a new approach?

ANIK: (*more subdued but still passionate*) I'm not saying I'm rejecting new ideas, but there's no need to reinvent the wheel here. My people—*our* people—need help. They're drowning in paperwork. Why not bring in some new people to help lighten the load? Maybe create a new position and a new department to take on specialized tasks!

Narrator: *Regina watches this exchange with equal parts interest and concern. She knows that conflict is inevitable with this group and that sometimes, it does produce great ideas, but it has to be handled properly or the meeting can turn nasty.*

REGINA: Look, it's clear that you both care deeply about your staff and want to find a solution. We're sitting down together to discuss this, and that's the first step. Let's zoom in on the one key issue that we all need to solve.

Narrator: *Regina opens her notepad and begins to take notes. There's something about a pen and paper that always helps her organize her thoughts.*

REGINA: We need to develop new services and deliver value to customers. That's the issue. How can we solve it?

SAMANTHA: Well, let's see! Other than those reports—and I realize we have so many of those—each of us has some autonomy within our division.

Over the years, as Royal Energy has become bigger, we've pushed this way of thinking down to the departments. Each of our functional heads has had near autonomy to improve their operations. In fact, Anik, your Marketing group runs things quite differently from my Marketing group! It's great to make things efficient and optimized for each department, but this separation is simply not helping.

Narrator: *Usually full of jokes, Anik doesn't realize how much his smooth demeanor is beginning to crack. His loyalty to his staff sometimes makes him defensive, and he struggles to reign in his passionate responses to what he perceives as Samantha's bulldozing commentary.*

ANIK: But that is literally how Royal Energy has been managed since the beginning . . .

SAMANTHA: Yes, that's the problem, isn't it? They aren't working together. *We* aren't working together.

ANIK: I don't recall your staff being overly helpful the last time we tried to work together.

SAMANTHA: Forgive me for saying so, but our approach to analyzing customer data was found to be more efficient than yours.

ANIK: I had a longtime employee quit over the disputes your staff created. I can't imagine they would be thrilled to work together again in any capacity. My marketing head does not want to work with your marketing head. Forget working with Finance or Sales—

SAMANTHA: And yet you're suggesting that we bring brand-new people into this internal political climate, assuming they would be able to handle their job on top of all our personal drama? I don't think so.

REGINA: Friends . . .

Narrator: *Samantha and Anik look at Regina, then look away. Regina cuts through the tension gently but firmly.*

REGINA: I didn't get a chance to speak to you both privately about that incident, but I'm going to now so we can put this matter to rest. It turns out that the conflict arose because the two workers involved knew each other personally. Well, add job hierarchy to the mix, and you've got a problem no matter where you go. In short, this wasn't your fault.

Narrator: *There is brief silence at the table. Regina wants to be the voice of reason and move the group past this stumbling block.*

REGINA: I know this has been a bone of contention between you two, but we won't be able to come up with a solution unless all of us work together. Are we agreed?

SAMANTHA: Yes, and thank you, Regina.

Narrator: *Samantha looks at Anik. She hopes that her expression is one of a truce, not defiance. Anik agrees as well, but hesitantly. Satisfied, Regina continues to take notes.*

REGINA: So, Marketing, Sales and Distribution, and other departments should work together more closely.

ANIK: Alright, I get it . . . but can we also add another option? We should create a new department called "Customer Service" and hire good people in that department. Besides, why is working across departments so damn hard? It seems so simple!

REGINA: Well, we're an established company. Like you said, we've done things the same way since we started. It's difficult to teach an old company new tricks.

SAMANTHA: But it can be done. We know that our customer base was smaller back then, but other than that . . . I think I know why what we were doing back then isn't working anymore. We've been asking each department, each function, to only manage their part. They barely know anything about how the other departments work because we basically taught them to mind their own business. So when a customer calls in with an issue, we haven't been able to sufficiently help them. We just keep passing them on to the next department.

ANIK: I have to admit, you're right . . . what you're saying is that if we keep doing that, we aren't really paying attention to the customer at all! We're saying "This is not my job—go talk to someone else in that other department." Agreed! That is a problem! How do we resolve this? Can we consider the second option we added to the list of ideas? A new department called "Customer Service"?

SAMANTHA: You really like that idea, huh?

ANIK: *(with a shrug)* It's a good idea, and you know I don't let a good idea die too soon.

SAMANTHA: Well answer me this: How is this "Department of Customer Service" going to come up with completely new product offerings like the renewable energy options our competitors are touting? We're not looking to just respond to customer complaints! And then, wouldn't the other departments just pass the customers on to the Customer

Service Department? They'll still not feel any responsibility to the customer. In fact, they'll just dig in even more into their own operations.

ANIK: Alright, I understand, but how would we even go about doing something that isn't tied to a Customer Service Department? You want all departments to understand what they're doing in relation to the customer, right? I mean, I wouldn't even know where to begin.

SAMANTHA: You know, I remember something I heard at the National Energy Conference in Dallas last month. A woman was talking about this problem of coordinating across departments. Were either of you there?

REGINA: No, what did we miss?

ANIK: I was, but I was so bored I don't remember the presenters.

SAMANTHA: Why do you even go?

ANIK: For the croissants.

SAMANTHA: Whatever. I can't remember her name, but she discussed how it can be difficult to think of designing and delivering new services to customers because it often requires coordinating across departments. She made a big deal of how these should be called processes. We can think of the entire company as made up of processes—not departments! And actually, Anik, the solution does involve bringing in people who can help with the coordination across functions. I think the role was called . . . process owner. And we need many process owners. Maybe that's the sort of thing we need.

ANIK: That's not what I was thinking of when I said "Customer Service Department"—I think if you're talking about improving customer service, then we have to make someone responsible for that! (*sighs*) Look, I don't know about having these process owners, if that's what you're calling them . . . but . . . okay. Let's play with this.

REGINA: This sounds promising, Samantha! Do you think you can find the name of the speaker and get us the material? We can review it and meet again in a week.

ANIK: I just want to say that we should not ditch the Customer Service Department idea yet.

REGINA: I hear your concerns, Anik, but I want to look at this idea Samantha's introducing more thoroughly.

SAMANTHA: Yes, I had my assistant move those presentation summaries into my drive. I'll go over them with her tomorrow and email the materials to you as soon as possible.

REGINA: Anik, how does this sound to you? Want to meet again next week after we've looked this over?

ANIK: Look, I've never even heard of this before . . .

SAMANTHA: Can we agree to at least look at the material before the next meeting?

Narrator: *Anik sighs, raising his hands in mock defeat.*

ANIK: Yes. I guess I can agree to that. But speaking of processes . . . this is all a lot to process right now. I don't know how the three of us are going to figure this out, especially with Mr. Rhode Island AWOL most of the time.

SAMANTHA: Yes, actually, where in the world is David this time?

REGINA: He's clearly missing out on what we've accomplished just by showing up today!

ANIK: Narrowly avoiding mediocre French pastries?

REGINA: We're working together! Solving problems as a team! I noticed that I had a problem within my regional center, and I figured maybe the two of you shared this problem . . . so, I invited you to collaborate. Perhaps we should do the same thing with our departments.

SAMANTHA: Like, contact people to work in groups like this and collaborate on ideas?

REGINA: Yes. I think this would be a great approach. We don't have to solve all the problems ourselves; we need other good thinkers to be in on it too.

ANIK: I can think of someone I'd like to include.

SAMANTHA: Matter of fact, yes . . . so can I.

REGINA: Great. Let's think of who we'd like to include and contact them. I'll type up the notes for this meeting and send them this afternoon for your records. (*she checks her watch*) I have another meeting across town in forty minutes, so I'm going to rush out. Good to see you both.

Narrator: *As Regina rushes to her next meeting, Samantha gathers her things to leave. Anik stands with his things ready to go. He wants to discuss the staff collaboration failure with Samantha, but his pride makes it awkward.*

ANIK: It's, uh . . . (*chuckles*) it's wild about that issue with our staff conflict thing, right?

SAMANTHA: (*smiles*) Yes, actually . . . I had no idea that was going on.

Narrator: *Samantha decides to bite the bullet.*

SAMANTHA: Look, Anik . . .

ANIK: Sam, I . . . I'm sorry. About what happened with the staff, my attitude during this meeting and before . . . I apologize. My ego got in the way.

SAMANTHA: Hey, it happens. For the record, I'm sorry too.

ANIK: I don't know how we're going to solve all of these issues . . .

SAMANTHA: (*shrugs*) We'll do it together. As a team.

ANIK: Aww, how touching.

To be continued . . .

Reflection Questions

1. To address the problems at Royal Energy, here's what Anik suggests: "There's no need to reinvent the wheel. . . . Why not bring in some new people to help lighten the load? Maybe create a new position and a new department to take on specialized tasks." Samantha disagrees. What's her counterargument? Whose argument do you agree with? Why?

2. Which path do you think is easier to put into practice? Samantha's or Anik's? Why?

3. What do you think Royal Energy should do? If you were asked to stand up and defend your recommendation in front of the VPs of the four divisions, what arguments would you make? Would your arguments change if you were asked to talk to the different department heads, such as Marketing, Operations, or others?

4. What obstacles do you anticipate Royal Energy will face if they follow the path you suggest?

Self-Assessment Questions

1. Can you describe what is a business process and what is not a business process?

2. Can you identify the difference between a business process and a function?

3. Can you explain why business organizations are often structured with functions?

4. Can you convince a colleague of the importance of focusing on business processes?

Readings

1. McCormack, K. 2001. "Business Process Orientation: Do You Have It?" *Quality Progress* 34 (1): 51–60.

2. Hammer, M., and S. Stanton. 1999. "How Process Enterprises Really Work." *Harvard Business Review* 77:108–20.

Key Terms

Actor
An actor is a person, group, or entity that performs a *task*. An actor can also be a device, an automated system, or a software application. Essentially, it is any resource that is capable of and responsible for carrying out a *task*.

Cross-Functional
The term *cross-functional* refers to anything (e.g., a *process*) that crosses functional boundaries. For instance, the order fulfillment process may involve *tasks* done by *actors* in Sales, Customer Service, Logistics, and Finance. Such processes require breaking *functional silos*.

Function

A function describes a grouping of capabilities often mapped to a department, such as Sales, Finance, and others. Different functions have the expertise to carry out different tasks—for example, Sales may execute tasks such as prospecting clients, and Finance may carry out tasks such as creating promotional materials.

Functional Silo

A functional silo refers to a department or *function* that focuses primarily/only on its own tasks and operates separately from others. The term is often used in a pejorative sense because it suggests a lack of coordination and communication between different departments.

Process

A process is a defined sequence of *tasks* carried out by *actors* to achieve a goal. Typically, a process is repeatable, with each instance of the process following the same series of steps.

Task (a.k.a. Activity)

A task describes a specific unit of work or activity, a step within a process, performed by an *actor* to produce specific outputs. Examples of tasks include a simple action like entering customer data or a complex task such as developing a sales forecast.

Scene 2
So Many Processes

An organization is seen, not structured into departmental silos
But as several processes that describe how the work flows

Learning Objectives

- Describe the phases in business process management (BPM)
- Explain possible approaches for process identification (the first phase in BPM)
- Assess the appropriateness of process identification approaches in different situations
- Distinguish "business functions" from "case types"

Characters

Kevin Sanders—Scheduling/Dispatch, New Hampshire
Jamie Cochrine—Operations, Rhode Island
Claudia Narez—Environmental, Eastern MA
Abra McGregor—Finance, Western MA

Narrator: *Two weeks ago, VPs from three of their four regional centers—Eastern MA, Western MA, and New Hampshire—met and discussed a key problem. They discussed possibilities such as a Customer Service Department and collaboration across departments, which they called processes. They promised to do some homework. Some phone calls later, they have now each sent a representative from their region to continue the discussion and develop the "process" idea. Sitting around the table in a conference room today are four rising stars representing their divisions, with the full endorsement of the division heads. We pick up the conversation after the early introductions...*

KEVIN: Okay, so, Claudia, I have to ask . . . How is it working with Regina Wood? I've heard she can be a bit of an enigma, you know?

CLAUDIA: What? No, I've actually found that what comes across as intimidating to some people is just how she pays attention—

KEVIN: Oh, like she's zoning out?

CLAUDIA: No, like she's—

KEVIN: Funny that a gal in that position wouldn't have learned better listening skills.

CLAUDIA: She's an excellent listener. That's why she has that job; she actually listens to people.

KEVIN: Oh, well, why'd you say she wasn't good at it?

CLAUDIA: (*bristling*) I didn't—

KEVIN: Just teasing.

Narrator: *Kevin winks at Claudia, but she just stares back at him. Claudia is here today because Regina knows that her levelheadedness can find a solution even in a tense situation. As it happens, the meeting was already starting out on the wrong foot. Kevin pulls out his cell phone to check the time. Claudia turns to Abra to see if she has witnessed the exchange. Abra is typing on her laptop with lightning speed, seemingly oblivious. Claudia wonders briefly if anything will come of this meeting. Just then, Jamie bursts into the room.*

JAMIE: Good morning. I am so sorry I'm late. I don't even have an excuse or anything, just extremely sorry.

Narrator: *Jamie Cochrane, manager of the Rhode Island branch's Operations Department, is the youngest at the table that morning. Coming up with new ideas is the reason he's been brought on the team in the first place. Kevin stands and puts out a hand to introduce himself.*

KEVIN: Don't worry about it, and don't apologize. Kevin Sanders, Dispatch, New Hampshire region. And, a word of advice, just thank people for their patience. No apologies necessary.

Narrator: *Kevin claps Jamie on the back in a fatherly way.*

JAMIE: Ah, okay . . .

Narrator: *Claudia stands and approaches Jamie.*

CLAUDIA: Hi, I'm Claudia Narez. Environmental, Eastern Massachusetts region.

Narrator: *They all turn to Abra, who is still typing on her laptop. Sensing their attention, Abra turns long enough to introduce herself.*

ABRA: Abra McGregor. Western Massachusetts, Finance.

Narrator: *Abra turns back to her typing, leaving a brief but awkward silence behind her. Making an executive decision, Claudia pulls out her notes and invites the others to take a seat.*

CLAUDIA: Okay then . . . let's sit down and figure this out. As you all know, we're exploring this idea of business processes because our operations are suffering—the customers are telling us they are not being heard and we're losing our edge. Yes, the business we are in is different in that we are a utility, but we cannot continue to do this. Here's what I am being told: Apparently, we aren't *connected* across functions, whatever that means. And because of this, our customers are suffering—

KEVIN: Yeah, I read that in the brief, and honestly? I don't know what to make of that. I mean, aren't we all supposed to do our jobs well? Isn't that what makes our customers happy?

CLAUDIA: Anyway, that is what we have in front of us. We have to think of better ways to help us all work together across departments—

KEVIN: Yeah, it's been awful on the ground for my guys. They show up to repair or install a service and the customer starts peppering them with questions about how they signed up and billing because (*laughs*) they can't get a hold of Finance! It's a nightmare.

Narrator: *Claudia expels a long breath as Kevin speaks, annoyed at being interrupted so many times. At the mention of Finance, Abra briefly looks up from her typing to give an unreadable glance toward Kevin and just as quickly looks back to her screen. Claudia notices Abra's glance and decides to try to remedy a possible perceived slight.*

CLAUDIA: Ah, yes, I empathize. My department has some big thinkers who would be on the rise if we had time and space to support new ideas. But I feel like the way the company is currently being run is holding us back. I'm sure Finance has been having similar issues.

Narrator: *Gesturing to Abra, Claudia waits for her response. After taking a deep breath to pull herself away from her notes, Abra looks up at the group.*

ABRA: Yes, it's been difficult.

Narrator: *Abra turns back to her computer. Claudia is puzzled. Kevin laughs uncomfortably and gives a pointed look to Claudia and Jamie. As much as the other two don't want to join Kevin's knowing look, they can't help it. The other divisions often see Finance as being a bit "above it all" when it comes to solving problems. Since Finance doesn't bring in any money, only managing what is already there, divisions like Marketing and Sales often have a complex about Finance. Confused by Abra's apparent disinterest in participating, Jamie seizes the opportunity to speak up.*

JAMIE: (*clears throat*) Operations has wanted to start using some of the ideas heard at the National Energy Conference a couple months ago. Well, I should say . . . *I* have wanted to start using new ideas.

	But I feel like my staff is so locked into the way things have always been. I'm wary about uprooting the system.
KEVIN:	Son, you've gotta be a man and just do what you need to do. You're their boss! Tell 'em you'll fire 'em if they don't do what you say. (*laughs at his own boldness*) I guess that's a little extreme, but you'll understand what I mean in time.
JAMIE:	(*slightly defensive*) I can't just fire people who have been working at the company for years just because they don't like change. Nobody likes change.
KEVIN:	Not true. I love change. The faster the better.
CLAUDIA:	(*cutting in*) I think this is the issue. We need to start making changes now because our customers are suffering. In our attempts to keep things organized and standard on our end, we are ignoring our customers. For example, there's a guy I know in Finance who does credit checks, billing statements, etcetera . . . he does his job really well and has recently been rewarded for his work, but he has no clue how his job fits into the overall process of delivering energy to the customer. So if there's a delay or something abnormal happens, the customer suffers and our star employee in Finance couldn't care less.
KEVIN:	Well, the way the organization regulates departmental tasks is outdated . . . I mean, we need it, but we also have to make our employees understand that what they're doing is eventually for the customer. And it seems that the old heads don't want to change. (*noticing that Claudia is about to speak*) Now, now, I know they *say* they want change, but try implementing it! They talk the talk but don't walk the walk, that's all I'm sayin'.
CLAUDIA:	Actually, I agree with everything you're saying. And I think we can also agree that we have to find a way to be more responsible to our customers, but we have to try to do it without flipping the whole boat over.
KEVIN:	I'm just saying, an overhaul is what we might have to do to make something work.
JAMIE:	I think we have the opportunity to see things from a customer perspective as well. Instead of worrying about how things are on our end, what if we started with the customers and worked backward?
CLAUDIA:	(*thinking*) I don't know, but let's add it to the list of ways to think about this. Remind the group about it later. Thank you, Jamie.
KEVIN:	Claudia, did you happen to come up with any ideas on how to not "flip the whole boat over"?

CLAUDIA:	Yes. I think we should make a list of things we know and things we need to figure out.
ABRA:	(*suddenly*) In conclusion . . .
Narrator:	*They all jump as Abra suddenly begins reading her notes off her laptop. She's been silent for so long, they've almost forgotten she's present.*
ABRA:	. . . we need to figure out what processes we have, how to find them, and how to name them.
	(*pause*)
CLAUDIA:	Yes, I agree that—
ABRA:	We also need to understand that there are going to be some processes that are unique to each division, but there are some that are going to be similar. So if we find and name them in Eastern Mass, they should work in Rhode Island—with minimal changes, we hope.
	(*pause*)
ABRA:	So . . . how should we proceed?
Narrator:	*Abra looks at the group expectedly, then turns back to her laptop, poised to type. The group now understands the way Abra wants to participate and communicate in the discussion. The conversation flows a bit easier.*
KEVIN:	Okay, so . . . lemme get this straight. We have our different departments like Finance, Sales, and such. Aren't those the processes?
CLAUDIA:	No, departments are doing specific things. Abra was explaining that processes need us to chain those things across departments.
KEVIN:	Wait, wait . . . I get it! Chaining things across departments . . . across functions . . . cross-functional.
CLAUDIA:	Yes, exactly. Thanks for bringing us full circle to the start of the meeting. And the word *cross-functional* makes perfect sense. Let's keep using it.
KEVIN:	So what you're saying is that each of our departments is specialized to do one set of things, and that means we have to get them to coordinate with one another. I get it. I'm telling ya, I'm a smart cookie too!
JAMIE:	Okay, but . . . if we see departments in our organization chart and do not see processes, that makes them essentially . . . invisible? Is that it?
CLAUDIA:	Maybe that's why the VPs want us to identify them. Right now, they're invisible. We have to find a way to bring them into the light, so to speak.

JAMIE: Okay, that makes sense. But even if we identify these processes, they run across departments, right? So who is responsible for each of these . . . processes?

CLAUDIA: I think we should focus on identifying the processes first and worry about who is responsible later.

KEVIN: Well, we know we're not the only energy company out here. There have to be other companies who have thought about this. There might be a, whaddya call it . . . a catalog or something of these processes we can use. I don't think we should have to make our own.

CLAUDIA: Royal Energy has a catalog, but it was last updated in the late nineties.

KEVIN: Geez, I've been working here for years, and I had no idea there even was a catalog. How did you find it?

CLAUDIA: I did some research ahead of this meeting. To my surprise, but apparently nobody else's, no one looks at the catalog at all anymore.

KEVIN: So, close but no cigar?

CLAUDIA: Well, I don't know. Maybe we can start there and figure out what is missing or what is obsolete. You know, I was wondering this morning what this meeting was going to be all about, but I'm beginning to see why what we're discussing is so important.

JAMIE: Processes . . . this reminds me of something. I just moved with my partner to be closer to the Rhode Island regional center—

KEVIN: Oh, I didn't know you were married. What does she do?

JAMIE: Ah, well, we're not married. And they don't identify with the gender binary.

KEVIN: Oh . . . (*doesn't understand*)

CLAUDIA: (*clears her throat*)

JAMIE: Oh yes. So anyway, we just moved there, and we had to set up our gas. Our real estate agent actually sent us a link to a website that would walk us through setting up our utilities in one place instead of having to go through each individual company ourselves. The website had a list of companies that serviced our area for electric, wireless and cable, and gas. Of course, we've all done these things ourselves, but this process was much more streamlined.

CLAUDIA: Abra, can you make a list of—

ABRA:	Already did. (*she begins to read aloud*) To initiate utility services at an apartment or house, the process begins when a customer requests the services. First, the request is recorded with details. Second, a qualified person examines the—
JAMIE:	Or software.
ABRA:	(*eyeballing Jamie*) Or software . . . examines the customer's location to see if it has been wired for service. Third, a different individual . . . or a software . . . looks up the customer's creditworthiness. Fourth, another individual or software—
KEVIN:	Jeez Louise!
Narrator:	*Abra stops reading. Everyone looks at Kevin.*
KEVIN:	(*changing his tone*) Thanks . . . for the work you did on this, Abra. I mean, gosh, I don't even know how you typed all this up so fast . . . and I don't mean to be rude, but can we get down to the brass tacks of this, please?
ABRA:	Certainly. Basically, there are many steps. Options are presented and responses are gathered about levels of service, scheduling, etcetera. A technician is scheduled and a physical device might be installed, like a router for wireless service, and then the service is activated by yet another person after the payment plan is set up by the person in Finance.
JAMIE:	Phew. I didn't even know there were that many steps, and I've been through it recently.
KEVIN:	That's a lot to follow . . .
CLAUDIA:	It is, but I'm glad we have a mind here who can put it all down. Thanks, Abra.
KEVIN:	So each of these tasks is done by a specialist. Someone from Finance is responsible for the credit check . . .
JAMIE:	Someone from scheduling is responsible for the technician's visit . . .
KEVIN:	Right, each individual is responsible for one set of tasks . . . it's like their "function."
CLAUDIA:	Exactly. And the tasks strung together across these functions become the process for initiating utility service for a customer. But the average employee in any given department is so specialized to their own function that they rarely get to appreciate the whole perspective.
JAMIE:	And the customer isn't even thinking about all this; they just want the final outcome. They don't know what goes on behind the scenes. Wasn't someone saying something about a catalog?

CLAUDIA: We were, but I remembered the old one was something like a thousand pages long. I imagine a new one would be even longer.

KEVIN: Yeah, knowing what we know about our work, this is gonna be some heavy stuff. Plus reading it and making sense of every aspect? That sort of goes against our goal to make this as streamlined as possible.

ABRA: (*still typing*) Jamie, didn't you say you had some ideas from an energy conference you attended?

JAMIE: (*surprised*) Yes? Yes. I did, I do!

             (*silence*)

CLAUDIA: What are they, Jamie?

JAMIE: Yes, yeah, so I was at the expo, and there were a couple vendors who had some standard processes implemented into a software.

ABRA: TBQ stood out to me.

JAMIE: Oh, I thought Sonita had a better approach, but—wait, were you at the same expo?

ABRA: Yes.

JAMIE: Oh. Well then why didn't you just say—

CLAUDIA: (*feeling weary*) Jamie.

JAMIE: TBQ and Sonita, yes. They're called ERP systems.

KEVIN: What the heck are those?

CLAUDIA: I believe it's enterprise software.

KEVIN: Oh, I've heard of this. I think in an article or something? I did that millennial thing of only reading the first few sentences (*everyone chuckles, even Jamie*), but I think I remember something about that. How would we use it?

JAMIE: The software? We'd just . . . buy it and roll it out, I guess.

KEVIN: Well, yeah, but people have to use it, you know. And like you said, the "old heads" won't take kindly to just rolling stuff out like that.

JAMIE: I don't think I've ever said the phrase "old heads" in my life.

KEVIN: Oh. Well, it's funny. Must've been me, hehe.

JAMIE: Ha-ha. So anyway . . . I agree, "stuff" would be difficult . . . changing the way people work and all.

ABRA: So buying a catalog or revamping the old one might be challeng-ing because it's too much work and would take too long. Buying software and rolling it out might be tough for some of our seasoned workers . . . I'm wondering if there's another way.

(*the group ponders the discussion*)

CLAUDIA: What if we did it ourselves?

JAMIE: What do you mean?

CLAUDIA: What if we came up with our own set of processes?

KEVIN: Didn't we already say that making our own catalog would be too difficult?

CLAUDIA: It would be hard, but it would be just as much work as cataloging or rolling out new software. Hopefully the end result would come more efficiently and be much better. I wonder if someone out there has figured out how to do this in a systematic way . . .

JAMIE: We could ask the employees what they would be more comfortable doing.

KEVIN: Right, but as we found before, a part of the problem is that our employees don't have a whole view of a process. So each need would be individualized, and we'd get a million different responses.

CLAUDIA: Hmm . . . so then . . .

ABRA: I have an idea.

KEVIN: Oh thank goodness, the brains of the operation. (*pause*) I . . . I'm gonna stop making jokes.

ABRA: We want to know how we use different functions together as part of a process, right? We need these individual employees to know the big picture and work together better. I believe our response should be to pull together different capabilities from different departments . . . effectively, like a process. We also have some options on how to start doing this.

CLAUDIA: What do you mean?

Narrator: *Abra begins to speak, but Jamie cuts in.*

JAMIE: I'm sorry to interrupt, but I feel like I need to remind the group that while we're trying to put more focus on what the customers need, there are also different types of customers. Like, there are residential customers, like people who live in my apartment complex, but then there are commercial customers, like Best Buy or Target chains. Not to mention schools and universities.

CLAUDIA: Right, and each of these different types of customers will have different types of requests. Jamie told us about how he set up his new service, but there are also issues like service interruption or cancel-ing a service. How can we figure out what processes are needed to respond to all the different requests?

ABRA: This is exactly what I mean. What if we think like that—pulling together capabilities—for all different types of customer requests? Then we can think of what capabilities we have to pull from the different departments for each type of request.

KEVIN: Woah, wait . . . that sounds like way too much work, and, I mean, who has that kind of information? I know I don't have access to that with my ground crew; we basically only deal with our own department.

CLAUDIA: Isn't that the point? We have to go past our own departments.

JAMIE: So actually, Abra's idea is an option. Like I mentioned at the begin-ning of the meeting, we can start with different types of customers and see which capabilities we need to pull from different depart-ments to respond to each type of customer. Honestly, that makes sense.

KEVIN: (*rubs face to clear his head*) So . . . is this how we want to roll it out, top down? Or should we try bottom up? I'm not making a joke this time; I actually want to know.

CLAUDIA: Wait, wait, what do you mean by "top down" and "bottom up," then?

KEVIN: You know, like . . . do we start with the customer or do we start with the employees? It's like . . . like, if you have a bucket of mushrooms, right? And there are all these different kinds of mushrooms in the bucket—

CLAUDIA: Mushrooms? Kevin, I already know this analogy isn't going to make any sense.

JAMIE: Wait, Kevin, I think I know what you mean. We want to figure out what processes we have and how to identify them. It's sort of like if we asked the question, "How does a laptop work?" Well, we could buy a laptop and then take it apart piece by piece. That's starting from the top down. Or we can get all the individual components and put them together to make a laptop. That's bottom up.

KEVIN: Right. Starting with the laptop and taking it apart is starting with the customer and considering their requests. Starting with the pieces is like starting with what we can do at Royal Energy and pulling things together across the departments—and building from there.

KEVIN: Hmm . . . I like this idea. But also . . . (*thinking*) that means we have to do both. The only question is where do we start—with the customers or our departments? I know I'm totally jumbled up here, but this discussion is starting to connect the dots for me, you know?

CLAUDIA: We can start a big catalog and we could buy some software. We have options. *However* . . . it's still all too much, and we really need this to be organized somewhere.

JAMIE: Well, I can agree on that much.

CLAUDIA: Same here. Abra?

ABRA: On it. Basically, what we have said is that we have three options. We can buy a thousand-page catalog and read through that to identify what processes we want—

CLAUDIA: I know it sounds labored, but lemme get someone to figure out where to get that just in case.

KEVIN: And whether it's even feasible in terms of cost and effort, you know? Sounds like a nightmare to me.

CLAUDIA: I know, but it would probably go over well with the longtimers because it would include things they're familiar with. Plus decades of industry expertise specific to this company.

ABRA: Or we can get a vendor to show us their software with standard processes and roll that out.

JAMIE: I'll put someone in contact with the company I met and compare different options with that. It would be expensive and rigid; we wouldn't be able to customize it at all. But it would come ready to roll out right away.

ABRA: Lastly, we can identify our own processes by looking at different customer types and seeing what capabilities we must pull together to respond to each person.

KEVIN: I'm still sort of confused about how we would do that. It's clearly a lot of work. Might be worthwhile but needs someone who knows what's going on.

ABRA: Kevin, would you like to work with my department to find an answer for this? Your joyful skepticism and my measured enthusiasm might make for more balanced results.

KEVIN: Ah, well, when you put it *that* way . . .

ABRA: Great. I'll have my people contact yours.

CLAUDIA: Well, it looks like we have our work cut out for us. But if we do this well, we should end up with a list of processes. Abra, we'll then need to come to you again for your finance expertise.

ABRA: Makes sense . . . if we have a portfolio of processes, we'll have to figure out what to do with it. (*suddenly perks up with an idea*) With a portfolio, we can figure out which processes to focus on first versus others, and so on, and so on . . . (*she begins furiously typing*)

CLAUDIA: This will eventually mean more work for you at this step. Is this workload okay for you?

ABRA: (*typing away, mumbling*) It's what I live for . . .

JAMIE: Well, at least we have options now. Good grief!

KEVIN: Dang, we deserve a raise! (*everyone laughs; even Abra cracks a smile*) I'm gonna go get one of these office kombuchas they put in the breakroom. Anyone wanna come? They're apparently really great for, ah . . . digestion. If you know what I mean . . .

CLAUDIA: As committed as I am to gut health, I have to pick up my toddler from preschool. Some other time.

JAMIE: I'll come grab some with you. Abra?

(*they all turn to Abra, still typing at her computer*)

ABRA: (*without looking up*) Nice to meet you all see you next time.

KEVIN: (*as they walk out*) Isn't she a little creepy?

To be continued . . .

Appendix 2A. Departments and Subdepartments/ Groups at Royal Energy

Sales and Marketing
Campaigns
Assessment
Retail
Commercial Accounts

Operations
Scheduling
Installation
Maintenance
Asset Analytics

Finance
Credit
Investments
Asset Management

Environmental
Health and Safety
Monitoring

Human Resources
Benefits
Recruiting

Billing
Subscription Management
Residential Billing
Commercial Billing
Revenue Management

Customer Service
Call Center
Analytics

Accounting
Receivables
Payments
Collections

Purchasing
Procurement
Receiving

Warehousing
Receiving
Device Stores

Research and Development
Analytics
Devices
Grid Management
Channels
Energy Sources

Technology Services
Web Portal
Intelligent Delivery

Other Departments
And subgroup . . .

Appendix 2B. Customer Types and Request Types[*] at Royal Energy

Industrial
  Installation
  Upgrades
  Repairs

Residential
  Installation
  Repairs

New Constructions
  Installation
  Repairs

Retail
  Installation
  Repairs

Institutional
  Contracts
  Installation
  Repairs

Partner Utilities
  Contracts
  Exchanges

---

[*] Drawing on Lucy (2011).

Reflection Questions

1. In the early part of the scenario, Abra launches into a longish description of what a utility company (such as an internet service provider) must do behind the scenes to accomplish what a customer wants (the installation of a new service). Working with a colleague in class, identify another example where the request that comes in from a customer shows just the tip of the iceberg, and all the many things that a company must do describe the iceberg that lurks beneath the waterline. Can you describe this in all its gory detail?

2. At one point in the scenario, Jamie shares this nugget: "If we see departments in our organization chart and do not see processes, that makes them essentially . . . invisible?" A little later in the discussion, Abra seems to enthusiastically endorse this idea. What do you think of this epiphany? What challenges would Royal Energy have to address to move to manage processes? How might they overcome these challenges?

3. Jamie and Claudia float the following ideas during the discussion: "While we're try-ing to put more focus on what the customers need, there are also different types of customers" and "Each of these different types of customers will have different types of requests." How would this help? What exactly do you think Royal Energy should do to incorporate these ideas?

4. Some of the other suggestions from the group to identify the set of processes include buying a catalog of processes and acquiring software that has predesigned processes. What might be some pros and cons of these alternatives? What criteria would you suggest for Royal Energy to select one of these approaches? Show how your criteria may lead to different selections for, say, a tech start-up, a university, and a niche manufacturing company.

Self-Assessment Questions

1. Can you explain possible approaches for identifying processes in an organization?
2. Can you assess the appropriateness of these approaches in different situations?
3. Can you distinguish functions in an organization from different case types?

Readings

1. Dumas, M., M. La Rosa, J. Mendling, and H. A. Reijers. 2018a. "Introduction to Busi-ness Process Management." In *Fundamentals of Business Process Management*, 1–33. Heidelberg: Springer Berlin.

2. Dumas, M., M. La Rosa, J. Mendling, and H. A. Reijers. 2018b. "Process Identification." In *Fundamentals of Business Process Management*, 35–37. Heidelberg: Springer Berlin.

3. Lucy, J. 2011. "The Customer Pyramid." Electrical Wholesaling, October 1, 2011. https://bit.ly/customer-pyramid.

Key Terms

BPM (Business Process Management)
BPM refers to the practice of designing, analyzing, and improving business processes to increase their efficiency and effectiveness by aligning with the organization's strategic goals and enabling continuous improvement.

BPM Life Cycle
The BPM life cycle describes a series of stages that are conducted as part of the practice of *BPM*, such as discovery and identification, modeling, analysis, redesign, implementation, and monitoring for continuous improvement of processes.

BPR (Business Process Reengineering)
BPR is an approach (introduced around 1990) to bring about radical changes to processes to improve performance with a focus on the customer.

Process Analysis
Process analysis (third step in the *BPM life cycle*) involves examining the current process (model and performance) to compare against benchmarks, evaluate effectiveness and efficiency, and identify any problems to identify opportunities for improvement or optimization.

Process Discovery and Identification
Process discovery and identification (first step in the *BPM life cycle*) recognizes the current processes that exist. This involves discovering and naming all processes and prioritizing them for further effort. Sometimes, the results are described as process architecture/landscape.

Process Implementation and Rollout
Process implementation and rollout (fifth step in the *BPM life cycle*) involves transitioning the employees and the organization from the current process to the redesigned process. It may include new technology design/deployment, changes to employee roles, and change-management efforts.

Process Modeling
Process modeling (second step in the *BPM life cycle*) is the creation of a graphical representation of a business process. It involves identifying all details, such as *tasks* and *actors*. The model can be examined for understanding, analyzed, refined, and used as a basis for implementation.

Process Monitoring
Process monitoring (sixth step in the *BPM life cycle*) involves ongoing observation and tracking to ensure that the new process is working as expected, to identify any discrepancies that could negatively impact performance, and to make any necessary adjustments.

Process Redesign
Process redesign (fourth step in the *BPM life cycle*) is when an existing process is incrementally improved or radically redefined (based on the results of *process analysis*) to minimize inefficiencies, remove bottlenecks, reduce costs, and improve customer satisfaction.

Scene 3
Prioritizing Processes

All these processes—great, now we can tell them apart
But with so many to consider, where do we start?

Learning Objectives

- Explain the rationale behind the life cycle view of process management
- Generate the process architecture for an organization with the case/function approach
- Describe how the set of processes may be evaluated from a portfolio perspective

Characters

Kevin Sanders—Scheduling/Dispatch, New Hampshire
Jamie Cochrine—Operations, Rhode Island
Claudia Narez—Environmental, Eastern MA
Abra McGregor—Finance, Western MA

Narrator: *At the end of the last episode, our big thinkers had three directions to explore. To pursue these, each pulled together a team from their home division. Over several weeks, this team met to discuss the ideas, research details, and compile possible solutions. The team carefully considered the different ideas but settled on one that the team members fully endorsed and developed. As part of this effort, they identified the various types of customers and their requests. For each request, they identified different departments within Royal Energy that must participate in crafting a response. They organized this data into a table. As the group meets today, these big tables have been printed and displayed on the walls in the conference room.*

ABRA: *(inspecting the table)* So I count about thirty or so columns. Each shows me a customer type. I see residential customers in one, I see commercial customers in another. Hmm . . . I understand. So different customer types.

KEVIN: And these column headings are broken into multiple columns some-
            times. I'm reading this as different kinds of customer requests . . .

ABRA: (*continuing her inspection of the table*) There are some overlaps. I see
            installation requests that may come from different kinds of custom-
            ers. But then we need contracts for a different set of customers—I
            see the value in this.

JAMIE: Interesting! And these rows we see—each of these seems to be a
            department we have in our divisions. Wow, that's a lot of depart-
            ments. I didn't realize we had that many.

CLAUDIA: There are departments—and then we have different subgroups
            within each department.

JAMIE: Okay, so here we have almost . . . let's see . . . ten, ten, ten, ten,
            another ten, and a bit—so almost sixty rows. That means about
            sixty subgroups within each division. We seem to be more similar
            across the divisions than we thought!

ABRA: So the X in a cell tells us that this particular subgroup must partici-
            pate in a response to that trigger. So, for example, I am seeing that
            the "New Residential Development" customer requires participa-
            tion from all of these subgroups across departments who have to
            do many different things. (*speaking to herself*) So the columns are
            showing me customer types and different kinds of requests that
            come from them, and the rows are showing me the departments
            and subgroups—I understand.

JAMIE: (*joking*) You love this, don't you?

ABRA: I feel that this is acceptable and well documented, if that's what
            you're implying.

Narrator: *The table has several cells with colors. A set of cells with the same color
            outlines a cross-functional process. Different colors identify different
            processes. There are some judgment calls in this. See how the team rea-
            soned that risk assessment for residential customers and billing setup
            can go together (dark-gray cells) and how they identified a process called
            the "Residential Account Setup Process." They also reasoned that risk
            assessment for commercial accounts and new residential developments
            have similar characteristics (light-gray cells), such as multiple lines, large
            accounts, and so on. The group then identified a process called "Risk
            Assessment for Commercial Delivery." Such vertical and horizontal
            slicing has allowed them to identify many processes. Each is listed with
            a name and a short description.*

ABRA: We're on to something here.

KEVIN: (*lost in thought*) Hmm . . . hmm? What?

ABRA: I feel that we're starting to get down to the brass tacks of this, as you would say.

KEVIN: Yeah . . . I don't know.

ABRA: What do you mean?

KEVIN: I feel frustrated. Like, we've put in a lot of work so far. Not just us, but the teams we employed to collect this data. And it's organized and everything, but . . . I feel like it should be prettier than this somehow.

Narrator: *Overhearing this, Claudia responds.*

CLAUDIA: (*laughing*) What do you mean prettier?

KEVIN: (*laughs, then sighs*) Look, I knew this would be a lot of work. Heck, we all knew it would be work, but honestly? I'm not really looking forward to poring over all of this today. Frankly, I'm bored with this. Let's be done already.

CLAUDIA: I know what you mean. Getting this far has been a struggle and a half, for sure.

JAMIE: My team decided to go the same way, but I think we were more successful because . . .

Narrator: *The group shares the data they collected and how they acquired it.*

CLAUDIA: Honestly, I feel really good about what we've been able to collect so far. We're essentially looking at all of our issues right here on this chart. Remember that we felt overwhelmed when we had to figure out what the processes were—how to even find them and how to name them? *And* that there were going to be processes unique to each division? We have answers to those questions now.

JAMIE: Right, plus we had to find some that were similar across divisions.

CLAUDIA: And we've accomplished that! Two hundred and thirty-six processes collected, named, and organized.

KEVIN: Yeah, 236 . . . now, what do we do with them?

CLAUDIA: Well, wasn't our goal to improve service to our customers? Now we do that by improving these processes.

KEVIN: Yeah but, Claudia . . . 236. How long is this gonna take? We gotta look at each one of these processes in depth. This could take years . . . maybe even a decade. That's definitely not the streamlining I thought we were going for.

CLAUDIA: It's only been a couple months, Kevin—

KEVIN: Right, I know, I *know*—

CLAUDIA: We know these things take time! And we just started the meeting—why are you so negative right off the bat?

Narrator: *Abra interrupts, her voice cutting through Claudia and Kevin's frustrated conversation.*

ABRA: *Let's just say . . . for a moment . . . that time and resources are not concerns.*

KEVIN: But—

ABRA: Just for a moment. Please. Don't think about expense or time. How would we proceed?

Narrator: *Kevin sighs and rubs his face.*

KEVIN: Well, we have to dive right into each one, and we'd have to figure out how each one worked. Like really look at who does what for each process, then maybe, somehow . . . we could analyze what we find? Does that sound right?

JAMIE: But first we'd need to know the KPIs for that specific process, and then we can analyze how it's been working for us.

KEVIN: Okay, okay . . . that's a step.

ABRA: What else?

KEVIN: Well, then I guess we do what Claudia says and try to make it better.

JAMIE: But how do we know it's *getting* better? Do we come up with a monitoring system for KPIs?

CLAUDIA: Yes . . . like "Time to Delivery" and "Customer Satisfaction," things like that. Some of it can be automatic, and sometimes the customer can opt in to give feedback.

KEVIN: Yes. Okay, okay . . . I see it all coming together now.

              (*the energy in the room is feeling more positive*)

JAMIE: So then . . . after we've done this for one process, regardless of time and resources . . . have we now built a perfect, eternal process that will work forever?

CLAUDIA: What do you mean?

JAMIE: Will this single process be perfect for all of time?

Narrator: *The group pauses to think about this.*

CLAUDIA: No . . . no, it won't. I see what you mean.

JAMIE: Once we build a process and roll it out, it is going to become the old process. And someone, somewhere, sometime will have to do it all again.

CLAUDIA: Right, in ten years, another team will be sitting where we are now discussing how to revamp what we are designing right now. And who knows how many more processes there will be by that time . . .

JAMIE: So . . . it's a cycle. We'll have to keep doing this.

Narrator: *A cloud settles over the group, all except Abra. Kevin notices.*

KEVIN: Abra, I know you tend to be a little . . . aloof . . . but you don't seem to be the least bit bothered or surprised by this.

ABRA: I admit that we're facing a challenge, but I'm familiar with this sort of . . . recycling. I had an acting teacher back in the day—

Narrator: *Abra notices their looks of surprise.*

ABRA: Yes . . . I used to act. Anyway, on the first day of class, he told us, "All technique is crap. The acting techniques I teach you today will be obsolete within a decade, if not by tomorrow." So . . . hold on tightly, let go lightly. We're going to do a lot of work, and then we're going to do it all again. Or someone else will.

Narrator: *Abra returns to her typing.*

KEVIN: That's actually a really cool, uh . . . way of thinking, Abra. Thanks for sharing. But the truth is, we don't have unlimited resources or time or . . . acting teachers. I'm gonna pull an Abra and break all this down for us using my own words. Okay?

CLAUDIA: Alright.

JAMIE: Sounds good, yeah.

KEVIN: Okay, so I get it now, I think. Here's what we have to do to examine each process: First, we gotta do a deep dive into it. To fully understand it, yeah? Then we gotta see how it's doing with KPIs. Then somehow, we gotta come up with a new-and-improved process. Then we gotta roll that out in Royal Energy, and then *keep* monitoring the process with KPIs to make sure it's actually improved. *But then* . . . but then . . . someone's gotta repeat this whole thing in ten years.

ABRA: Because by then—or maybe sooner than that—some things may change so much that the process will need to be revamped. Yes, sounds about right.

KEVIN: And then we gotta do this for each process?

CLAUDIA:	Yes, we—
KEVIN:	Two hundred and thirty-six, and we gotta do it for all of them? And what we do may only stick for a few years? And . . .
CLAUDIA:	Kevin—
KEVIN:	(*getting frustrated*) And then someone's gotta just repeat this whole stupid thing!
CLAUDIA:	*Kevin!* This is not stupid. This is how we . . .
KEVIN:	Yes, but—
CLAUDIA:	This is how businesses are run! We keep getting better and improving how we operate. This shouldn't be—
KEVIN:	I *know* that, but 236—
CLAUDIA:	Kevin, we *know* how many there are, you don't have to keep—
JAMIE:	(*stepping in*) Okay! Okay. I think we need a break here.
Narrator:	*The group disperses. Claudia and Kevin show frustration but also embarrassment at their own outbursts. It's clear that they've both worked hard on this and care deeply about the project, but it's time for a break. Claudia grabs an apple from the table and walks out for some fresh air. Kevin pulls out his cell and starts playing a short game on his phone. After fifteen minutes, the group resumes the meeting.*
CLAUDIA:	Jamie, thanks for recommending the break. Kevin . . . I apologize—
KEVIN:	Hey, hey . . . no need. I'm the one who's sorry. I . . . (*sighs*) I just don't want all this work to be for nothing, you know? We've got teams of people working on this, and it's still nowhere near finished. I want to be optimistic, but this way of working never came easily to me.
ABRA:	Actually, the issues you raised, however arbitrary they seemed at the time, brought up some more specific questions in my mind.
KEVIN:	Gee, thanks, Abra . . .
ABRA:	Kevin was essentially asking, Where do we start? If there are so many processes, we know we cannot do all of these at the same time. I think we need to start by asking which one to attack first.
JAMIE:	We could have a team look at each one—
KEVIN:	Two hundred and thirty-six teams? I don't think so.
Narrator:	*Kevin catches himself.*
KEVIN:	I'm sorry. I guess I'm still stuck being the pessimist in this meeting.

CLAUDIA:	No, that's a good point. If we did something like that, all our operations would literally stop. We'd be in the same position we're in now, drowning in paperwork, but we'd have twice as much actual work.
ABRA:	So the word of the day for us seems to be *priority*. It comes down to prioritizing processes. How do we do that?
JAMIE:	Well . . . we could start by looking at the processes that cut across many departments. That would sort of be like killing multiple birds with one stone.
CLAUDIA:	That's a possibility . . . and a disturbing metaphor.
KEVIN:	We could always just do whatever our CEO would rather begin with.
CLAUDIA:	Daniela does have great ideas, but why would we want to simply figure out what the CEO wants? She assigned the VPs to this project, and they assigned us. Clearly they want us to figure things out on our own. Personally, I think we should focus on the processes that are the most expensive.
ABRA:	That would be difficult. Costing doesn't give us that data.
CLAUDIA:	Okay . . . well, maybe we can find out what matters to the customer.
JAMIE:	Yeah, but which customer? We have different processes for each customer.
KEVIN:	Well, one thing that our word of the day has taught me is that we don't have to deal with all these processes all at once. We can't, anyway, but we don't *have* to. That's promising!
CLAUDIA:	I think we should go back to the core question: How do we deliver value to our customers? We know we need to improve our processes, but we're stuck on which ones to start with.
ABRA:	Our choices are as follows: (A) We could start with processes that cut across many departments. (B) We could figure out what Daniela wants. (C) We could, somehow, do the most expensive ones first . . . or the least expensive. Or (D) we could look at what matters most to the customer. Which should it be?
Narrator:	*The group ponders the options but cannot come to a consensus on which criterion they should use to prioritize. Why should the number of departments matter? If they want to truly improve operations, why should personal opinion matter so much? Cost considerations are important, but how can they get costing numbers for things that go across departments? Maybe they do need to look at what matters most to the customer, but is that enough? After staring at the chart for a while, Abra speaks up.*

ABRA: I think we need some early wins. Like Kevin said, we cannot consider all these processes at the same time, so maybe we need to first look for the processes that are not doing well.

CLAUDIA: Hmm . . . that's good, Abra, but how will we know which to choose?

ABRA: We can ask what the customers are most unhappy about. Looking at their complaints, perhaps.

CLAUDIA: So looking at what matters most to the customer—we have that already.

ABRA: Yes, but more than that—what is not working well for them. So if there are no complaints about the billing processes, we do not have to prioritize that. But if there are complaints about the install process, then we should prioritize that.

JAMIE: You know, Abra, sometimes it seems like you have all of the answers already . . . and you're just waiting for us to catch up to you. What gives?

ABRA: I like complexity. I hate easy answers. I live for the process of finding the answer to a difficult question. Although it feels like the best answers are the simple answers—however complex the question.

JAMIE: Like if you hear hoofbeats, think horses, not zebras.

ABRA: . . . Right.

KEVIN: So, I like that. But what about some of our internal processes? Like payroll and stuff.

ABRA: Yes, we have to look at those as well. But maybe we should prioritize processes that are external-facing instead of the internal processes.

JAMIE: What if we looked at processes that are, like, you know, more important ones?

KEVIN: Wait, wait . . . someone tell me, please—what makes a process important?

JAMIE: When it brings in money? So our billing process—even if it's not one that the customers complain about—is important to us, right? It's what brings in the money?

CLAUDIA: Okay, so we seem to have more criteria now. Remember we had four before. Let me add these two: (E) processes that customers complain about and (F) processes that bring in money. I am beginning to like these last two.

KEVIN: Sorry to break up the party, folks. I think it's time to wrap up for me. I like where we're going with this. Can someone get this summarized and share? And then we can decide which of these criteria we really want—yeah?

Narrator: *Everyone pauses and turns to Abra. She feels the gaze of the group and looks up from the computer.*

ABRA: Of course, happy to do that.

Narrator: *While not entirely satisfied with the results of this meeting, the group disperses knowing that the next steps they take, however small, will continue to lead them toward their goal.*

To be continued . . .

Appendix 3A. Identifying Processes at Royal Energy

Customer and Request Types	Industrial			Residential	
Departments and Groups	Installation	Upgrades	Repairs	Installation	Repairs
Sales and Marketing					
Campaigns	**1** X	X		X	X
Assessment	X	X	X	X	X
Retail					
Commercial Accounts	X				
Operations					
Scheduling	**2** X	X	X	**3** X	X
Installation	X			X	
Maintenance		**5** X	X		**7** X
Asset Analytics	**4** X	X	X	**6** X	X
Technology Services					
Web Portal	**8** X	X	X	**9** X	X
Intelligent Delivery		X	X	X	X
Finance					
Credit	X				X
Investments					
Asset Management		X	X	X	X
Environmental					
Health and Safety	X			X	
Monitoring				X	
Billing					
Subscription Management	X			X	
Residential Billing				X	
Commercial Billing	X				X
Revenue Management	**10** X	X	X	X	X
Accounting					
Receivables	X			X	
Collections	X	X	X	X	X
Payments					
Customer Service					
Call Center	**11** X	X	X		
Analytics	X	X	X	X	X
Warehousing					
Receiving					
Device Stores	X	X	X	X	

Customer and Request Types	Industrial			Residential	
Departments and Groups	Installation	Upgrades	Repairs	Installation	Repairs
Research and Development					
Analytics	X			X	
Devices		X		X	
Grid Management					
Channels		X			
Energy Sources		X			
Purchasing					
Procurement					
Receiving					
Human Resources					
Benefits					
Recruiting					

Notes

1. To create the table, the team first used the departments and groups as rows and the customer and request types as columns.

2. Then the team placed an *X* in the cells in each column to indicate that these sub-groups work together to craft a response to that request type from that customer type.

3. Finally, the team used rules of thumb* to draw boundaries around some cells, separating cell clusters, each describing a process.

4. The table illustrates only a few columns. The team created multiple tables, each showing a subset of columns mapped against all rows.

5. The table illustrates the identification of only a few processes with the boundaries. Fully populated tables were made available to everyone during the meeting.

---

* See Dijkman (2012).

Appendix 3B. List of Processes at Royal Energy

#	Process Name	Process Description
1	Marketing Campaign Rollout	This process coordinates the rollout of a marketing campaign, including its ongoing monitoring and assessment.
2	New Install, Industrial	This process coordinates the installation of a new service at a commercial or industrial location.
3	New Install, Residential	This process coordinates the installation of a new service at a residential location.
4	Invoicing Setup, Industrial	This process occurs after a new install to set up invoicing for a commercial or industrial client.
5	Maintenance and Analytics, Industrial	This process coordinates efforts for ongoing maintenance and analytics for industrial and commercial clients.
6	Billing Setup, Residential	This process occurs after a new install to set up invoicing for a residential client.
7	Repairs and Maintenance, Residential	This process coordinates efforts for ongoing maintenance or repairs, as needed, for residential clients.
8	Portal Access Provisioning, Industrial	This process coordinates efforts necessary to set up and provision portal access for industrial and commercial clients.
9	Portal Access Provisioning, Residential	This process coordinates efforts necessary to set up and provision portal access for residential clients.
10	Cash Flow Monitor Process	This process coordinates receivables and collection efforts toward revenue management goals.
11	Customer Service Process	This process coordinates work by the call center to respond to customer requests and complaints, drawing on and contributing to data captured and used by the analytics group.
. . .		
. . .		
236		

Notes

1. The table lists the processes, each with a name and a brief description of the process.

2. The table names and describes only a few processes. Fully populated tables were made available to everyone during the meeting.

Appendix 3C. Process Portfolio

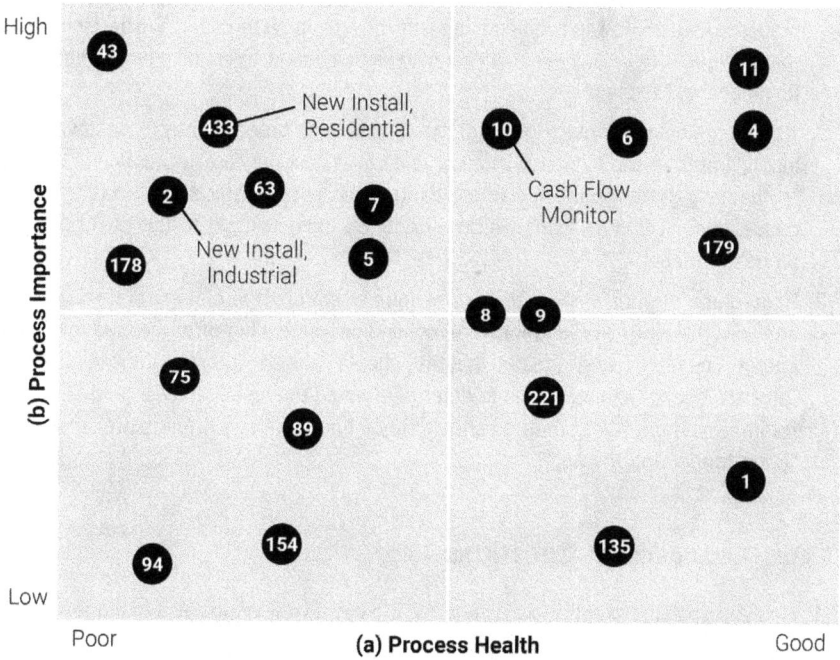

(a) Process Health

(b) Process Importance

Note

1. To generate the process portfolio, the team placed each process along two dimensions: (1) process health on the x-axis and (2) process importance on the y-axis.

Reflection Questions

1. Early in this scenario, we hear Claudia express her satisfaction upon looking at the chart: "Remember that we felt overwhelmed when we had to figure out what the processes were—how to even find them and how to name them? . . . We have answers to those questions now." What is your assessment? How will such a table/chart help Royal Energy? Explain.

2. A bit later, Kevin acknowledges that to examine each process, they must "dive deep into it," improve and roll it out, monitor to it make sure they are improved—and then, "someone's gotta repeat this whole thing in ten years." That sounds like a losing cause, doesn't it? Why can't we do a good job once and get to the best possible process for each?

3. Regardless of your answer to the previous question, it is clear that through this analysis, the team would identify, name, and describe a list of processes for Royal Energy. The team has suggested that they should consider "process importance" and "process health" as the criteria to determine which processes they should prioritize for improvement. What metrics should Royal Energy use to rate the processes for "importance" and "health"?

Self-Assessment Questions

1. Can you explain the rationale behind the life cycle view of process management?
2. Can you generate the process architecture with the case/function approach?
3. Can you describe how processes may be evaluated from a portfolio perspective?

Readings

1. Dijkman, R. M. 2012. *Designing a Process Architecture: A Concrete Approach*. Eindhoven University of Technology technical report. Direct link no longer extant.

2. Dijkman, R., I. Vanderfeesten, and H. A. Reijers. 2016. "Business Process Architectures: Overview, Comparison and Framework." *Enterprise Information Systems* 10 (2): 129–58.

3. Rosemann, M. 2006. "Process Portfolio Management." *BPTrends*, April 2006.

Key Terms

Case/Function Diagram or Matrix

A case/function diagram or matrix shows the different *case types* as columns and the different *functions* as rows. An *X* in a cell indicates that the function must be involved in providing a response to the case type. See the table in appendix 3A for an example.

Case Types (a.k.a. Customer/Request Types)

Case types describe the different types of customers/requests the organization receives. These often map to products/services offered by the organization. For instance, auto/home insurance claims represent different case types, as do simple/complex mortgage applications.

Function

See the definition as part of Scene 1: Thinking in Processes.

Process Identification Heuristics

The case/function matrix shows all mappings between *case types* and *functions*. One uses process identification heuristics to draw boundaries around certain cells. An example of a heuristic is a logical separation in time or space. See step 3 in Dijkman, Vanderfeesten, and Reijers (2016) for a list of heuristics. To see examples of processes identified by applying such heuristics, see the numbered rounded rectangles overlaid on the table in appendix 3A.

Working with a Process

Scene 4
Modeling a Process

Let's focus on one process, capture how it works
How? Do I reveal all details? Do I flaunt all quirks?

Learning Objectives

- Explain fundamental ideas underlying models of business processes
- Distinguish modeling approaches for business processes from other approaches
- Build a process model to reflect information from stakeholders

Characters

Kevin Sanders—Scheduling/Dispatch, New Hampshire
Jamie Cochrine—Operations, Rhode Island
Claudia Narez—Environmental, Eastern MA
Abra McGregor—Finance, Western MA

Narrator: *In the last episode, the group floated several possible criteria to prioritize processes. Since then, they have had more phone calls, enlisted help from their internal teams, and collected more data. Now they have something that shows each process in a two-dimensional space with 236 dots, one for each process. The team is meeting in a different conference room today. They are all examining the chart, except Jamie, who is blown away by the amazing view from this conference room.*

CLAUDIA: Okay, so can everybody see this? We're going to start with . . . (*notices Jamie staring out the window*) Jamie? We've started.

JAMIE: Sorry, geez, this view is so gorgeous. Abra, I had no idea you worked in such a cool building.

KEVIN: Right? There's a pool table downstairs. And a bar! It's like a WeWork, but exclusive to our company. Abra, think you can put me in for a transfer?

ABRA: The relaxed environment with accessible recreational options seems to improve work attitudes, but I do find them to be a distraction. Sort of like right now.

                (the guys get the hint)

JAMIE: Ah, yes.

KEVIN: Sorry 'bout that.

CLAUDIA: Okay, so just so you all know, this chart that I'm projecting from my laptop is interactive. We can add or take away processes as we see fit.

ABRA: From what I recall, we talked about processes that make money for us and processes where customers complain. Does this sound accurate?

CLAUDIA: Yes. We have value for the process importance on the vertical axis and process health on the horizontal axis.

JAMIE: And just to clarify, process importance is . . . ?

CLAUDIA: How critical the process is to Royal Energy. It may be because it is an external-facing, money-making process.

KEVIN: Or because it's something we gotta do to keep the regulators off our backs.

CLAUDIA: Exactly. And the process health dimension is what tells us whether our customers are complaining a lot or whether the process is alright for now—and customers here may be whoever uses the process, including regulators.

ABRA: We can see the full set of processes on this one chart. Sufficient.

KEVIN: But this is an interactive chart, yeah? Well, we'll have to do a lot of interacting because holy mother of the gods, there are a lot of dots.

CLAUDIA: Yes, and we all know how many there are, so we don't have to—

KEVIN: *(singing)* Two hunned and thirty-siiiiiix . . .

CLAUDIA: Oh boy. Okay, yes. And that's too many on this particular graph, so just for today, we've narrowed it down to eight.

Narrator: *Claudia clicks some keys, and 228 dots fade from the chart. This leaves 8 dots in the top-left quadrant.*

KEVIN:	Wait—why? Oh! (*he suddenly realizes what the chart is all about*)
ABRA:	So these are the processes that are critical. High on the vertical axis, but not doing so well, so low on the horizontal axis.
CLAUDIA:	Yes, process health is a concern for these based on customer complaints.
ABRA:	And that means we need to prioritize these. I can see where things are going.
CLAUDIA:	So even though we have these eight processes identified as high-priority processes, I think we need to figure out which one we want to attack today. Let me start by naming some of these. We've got maintenance, a couple of the new install processes, interconnect, cash flow management, and some others. Anyone have a preference on where we should start?
JAMIE:	I know a little about the maintenance process; I looked over Claudia's research on the old handbook. It was basically designed back in the 1980s to make sure there was a cyclical structure of continued maintenance. It covers everything from the monthly upkeep of the underground electrical wiring and protective tubing to how many times a week the bathrooms are cleaned at the Rhode Island branch.
KEVIN:	Yeah, we use a similar process where I am.
CLAUDIA:	We do too. Abra, I think you do too?
ABRA:	Yes, and actually, I don't think there's a pressing issue with this process. In fact, I'd say it works pretty well. We may see some complaints, but I'm guessing these are about convenience. We've actually not had any accidents we need to worry about, so I think we're good on this one—more or less.
CLAUDIA:	More or less?
ABRA:	Everything can always be improved upon, but this is not necessarily a crucial process at this moment and it's working well. So I vote that we don't prioritize it in spite of what our chart is pointing to.
Narrator:	*Everyone nods, and the group looks at the new install processes next. They decide that, although critical, it seems to be working well enough and doesn't need their full attention, at least not just yet—again, in spite of the customer complaint data. Next, they examine the cash flow management (CFM) process.*
KEVIN:	Oh boy, this one just *kills* me.
CLAUDIA:	Yes, this is a process I think we all equally despise.
JAMIE:	I'm pretty sure that we all handle our CFM differently. We can't even help each other out. I know Western MA recently hired a new

group of senior analysts to basically cover all of this, and I hear it's not going so well.

ABRA: You'd be right about that.

Narrator: *The four of them share stories about how much this process has caused them trouble and how it could be worthy of redesign. They also confirm that the process is essential to the company. However, the sense in the group is that the improvement would be at the margins. So although they can improve the process to be technically operational, the financial impact would be barely noticeable.*

JAMIE: Turns out that the annoyance makes a louder noise than fixing it would. Let's just table this one for now and see what we have next.

CLAUDIA: Okay, what about that interconnect process?

KEVIN: Yes. That one.

Narrator: *The others in the group turn to Kevin, who has just so adamantly made his choice.*

JAMIE: Care to divulge your reasoning, Kevin?

KEVIN: Yeah, I can speak from experience. This process is highly critical, and it absolutely *stinks*! From buying supplies from the right vendors to installation—gosh, even making sure that the technicians on the ground have the right tools for install—everything stinks. It's arguably the most frustrating part on the technical side.

CLAUDIA: Well, the research from our teams confirms your experience.

ABRA: I agree as well. We may have a candidate.

CLAUDIA: Now, if we can do a good job studying and redesigning this process, we could show a clear "win" to the company.

KEVIN: I'll say! We'd strike gold just by making the attempt. Anything has to be better than whatever the heck this process has going on right now.

ABRA: A "win" would be good for our project. We've been working on this for a few months now, and there's not much to show for it. And by that I mean nothing solid . . . other than research, charts, and a better understanding.

CLAUDIA: All of which are necessary—

ABRA: I'm not arguing against that. I'm actually in support of where we are now. I'm just saying it would be nice to be able to show some tangible progress. This chart and all this research tell us where to go, but now we need to go there.

CLAUDIA:	Sorry, Abra, I've been getting a little uptight and defensive about this project recently. We've done so much work, it's hard for me to understand when it seems to be under attack.
ABRA:	It's alright, let's just move on.
Narrator:	*Claudia's still a bit stung, but that's just how Abra handles conflict. Truth be told, Claudia's been carrying more responsibility for this project than the others, though nobody else seems to appreciate it. She is tired, and it seems that every new discovery just shows how much more work they still have to do.*
CLAUDIA:	Okay, so . . . interconnect process. We'll look at it in depth and try to improve it.
KEVIN:	Just so y'all know, this is gonna be long and complicated. It's not the easiest choice here for our first try.
JAMIE:	But fixing it will make a big impact. I'm willing to go all in, headfirst.
KEVIN:	Well, hold your nose; it's a long drop and the pool is deep.
CLAUDIA:	Kevin, can you give us a sense of where to begin with this process? Maybe dipping our toes in at the shallow end?
KEVIN:	(*sighs*) Ugh . . . the whole thing is so convoluted, I don't even know where to begin. I wish we could make it easy, like . . . like drawing a picture. Or something. I dunno . . .
ABRA:	That actually might be a good idea.
KEVIN:	What is?
ABRA:	Making pictures of processes. A flowchart. That might be a place to start.
KEVIN:	(*surprised and pleased*) Oh! Wait, wait . . . I remember those from way back. (*pauses*) But is that all this has been all along? It took us so long to just get to flowcharts?
JAMIE:	As long as we got here, right? I've made a few of these myself. There's fancy software we could use!
	(*Jamie and Abra speak at once*)
JAMIE/ABRA:	I could pull up some—
ABRA:	Oh. My apologies . . .
JAMIE:	No, no! Seems like we both have experience with this, Abra. Mind if we work together?
ABRA:	(*pause*) Nobody ever . . . *wants* to work with me . . . (*recovers*) I mean, I don't mind at all.

Narrator:	*Jamie takes a chair next to Abra and pulls out his laptop.*
JAMIE:	Oh, you've got the same software I do. Perhaps—
JAMIE/ABRA:	We could do a shared document and . . .
Narrator:	*Abra laughs, Jamie smiles. Claudia and Kevin look on, bemused. While the two tech experts—Abra and Jamie—coordinate to start a shared document using their software, the other two work out what to do next.*
CLAUDIA:	So . . . Kevin, do you have an idea of what we would need to include in the flowchart?
KEVIN:	Well, again, it's complicated. There are triggers, like customer requests, or there are things that people do one after the other just because . . . well, because that's the way it's always been done. Hmm . . .
CLAUDIA:	Can you think of what the most important parts are?
KEVIN:	Okay. The first thing is usually we receive a form from a customer. Then we—
CLAUDIA:	Wait, so . . . (*she begins writing on a pad*) the first step is that a form is submitted. So that means that the first step is that the customer has to fill out a form?
KEVIN:	Nah, I think we can assume that. We don't have to say "First find the form, then fill it out, then submit it." We wanna capture the essential task; it's a judgment call on the details, I think.
CLAUDIA:	Got it. Then what?
KEVIN:	Then we gotta figure out if they're on a secondary system, whether power will be exported, whether the equipment is certified, and on and on. There are different directions the flowchart can go for yes or no answers.
CLAUDIA:	Uh-huh . . .
KEVIN:	And honestly? I could keep listing things I know, but I'm bound to miss something. I mean, I know I'm the brains *and* the brawn of this group, but I don't know everything, ha-ha.
CLAUDIA:	Well, just tell me as much as you do know . . .
Narrator:	*Kevin and Claudia attempt to make a complete list of the process, but it turns out to be just as complicated as Kevin predicted.*
CLAUDIA:	Okay, so the capacity needs to be checked, but only once? How many times?

KEVIN:	Uh . . . two?
CLAUDIA:	You don't sound convinced. Is it more?
KEVIN:	Well, sometimes the guys . . . I mean, the technicians . . . just sort of wing it.
CLAUDIA:	"Winging it" isn't exactly something I can put on a chart, Kevin!
KEVIN:	(*throws hands up in defeat*) Well!
Narrator:	*Abra and Jamie rejoin the conversation.*
JAMIE:	Okay, we're set up to do at least a mock-up flowchart. Do we have the information we need?
CLAUDIA:	(*sighs with frustration*) No, not at all.
KEVIN:	Well, you're acting like this is my fault or something—
CLAUDIA:	No, Kevin, I'm sorry. Maybe today isn't my day . . . I guess I'm feeling the same way you did at our last meeting. I just want us to be able to get all the information we need in one meeting.
KEVIN:	I know, it's annoying. But in this case, we really can't.
JAMIE:	Why not?
ABRA:	Because we need to see the process from all sides, from every angle. Not just from Kevin's point of view but from the people who actually do this work.
KEVIN:	I hate to state the obvious, but I think Abra is right. I thought I knew how it all worked, but as we've been talking, I see that there are a lotta holes. We really need to talk to the different people and departments who do different parts of the process. Nobody knows how all the different steps work, and that includes me.
ABRA:	I completely agree. As we've worked on this project, I think it is quite clear that no single individual will know the entire process. In fact, different departments can only tell you what they do and maybe what happens just before and right after. We will have to compile this and somehow put it all together.
KEVIN:	Right. So I know where scheduling comes in, how the customer orders group tells us to coordinate the equipment and technicians, and what we do next to hand off to Finance—but I see what you're saying. Each department can only know a part of the complete process. Wow, this stuff is finally beginning to make some sense!
ABRA:	So to construct a picture of the process, we will have to gather a lot of information from a lot of places before we somehow construct the process model.

CLAUDIA: Okay . . . so, say we get that information somehow, and then we put it all into the flowchart.

JAMIE: Well, not so fast. We actually, well . . . Abra and I actually—

ABRA: A flowchart won't be enough to accommodate this information. It's a start, but it won't give us an entire picture of the process like we hoped. We need another format.

KEVIN: Wait . . . what? (*annoyed and worried now*) Why?

JAMIE/ABRA: Because . . .

JAMIE: (*laughs*) Oh, go ahead.

ABRA: Because a flowchart has a totally different perspective. It just doesn't have what we need. It sounded like a good plan, but flowcharts were really meant for programming software. Not for a business process.

CLAUDIA: Now I'm gonna pull a Kevin—my brain is about to explode.

JAMIE: Don't lose hope. It's actually a relief to know that Kevin was right . . . it's impossible for one person to know everything about a process. Yes, this means we probably need to get teams together again, but at least we know we can't do it by ourselves.

KEVIN: Right! Many hands . . .

ABRA: Are you going to finish the saying?

KEVIN: No, we all know what it says. It's . . . it's funny. I'm funny! But I also don't see why we have to stop doing this as a flowchart . . .

CLAUDIA: Okay, moving on. What's next?

KEVIN: Let's gather people from different departments to work on this process. Figure out who does what.

JAMIE: Yes, we'll be looking at different documents, different data, and like Kevin said, different roles in the process.

CLAUDIA: And we won't get a final picture from all of them. We have to sort of stitch all the data together to form the picture. Should we do this manually, or . . . ?

ABRA: Leave it to me; I think I have an idea, and I'll work on it for our next meeting.

CLAUDIA: So I have to say that I am a little perplexed. Why are we not doing this as a flowchart? I know we talked about it, but I find myself in a strangely unhappy place. Why not a flowchart?

KEVIN: Yeah, I had the same question.

ABRA: So maybe we need to ask a simple question to ourselves: What do we want in a picture, a model of a process? I feel that we need to show what triggers a process, we need to show the different tasks in the process, we need to show who does each task—what else?

CLAUDIA: (*warming up to the idea a bit*) Yes, yes, and we may want to show how long each task takes and maybe what information is needed to complete the task.

ABRA: And we may need to show things like how a process may be different when a customer has a previous billing account, or sometimes, when two departments can do different tasks in the process simultaneously.

CLAUDIA: Okay, okay, you have me convinced. A simple flowchart may not be enough. But—is there, like, a standard way to come up with pictures or, as you say, models of processes?

KEVIN: And again, can't we just do all this with a flowchart?

ABRA: I'm not sure. There may be . . . with so many apps and so many software companies and consultancy companies, I wouldn't be surprised if we have some options.

KEVIN: (*looks at the watch, resigned*) Okay, fine, let's contact our teams and ask them to look for options. Is that what we want?

               (*suddenly everyone realizes that it is time; the meeting has flown by*)

CLAUDIA: I think so. Let's hope we can move this along. Great work today, everyone.

Narrator: *On the other side of the table, Jamie decides to make a new friend.*

JAMIE: Abra, I was actually hoping, well . . . if it wouldn't be too much of a—

ABRA: (*still typing at her computer*) Hmm? What?

JAMIE: (*gathering his nerve*) The new picture we're putting together for the data, I want to work on it with you. I mean, I would like to . . . if you, I mean . . . if I can help—

ABRA: I just sent you an email outlining my plan for moving forward.

JAMIE: (*phone dings*) Oh, I didn't know—oh! Okay, I—

ABRA: Give me your thoughts by tomorrow?

JAMIE: Yeah, yes! Yes. I can. Thanks. Bye.

To be continued . . .

Appendix 4A. A Model of the Interconnect Process at Royal Energy

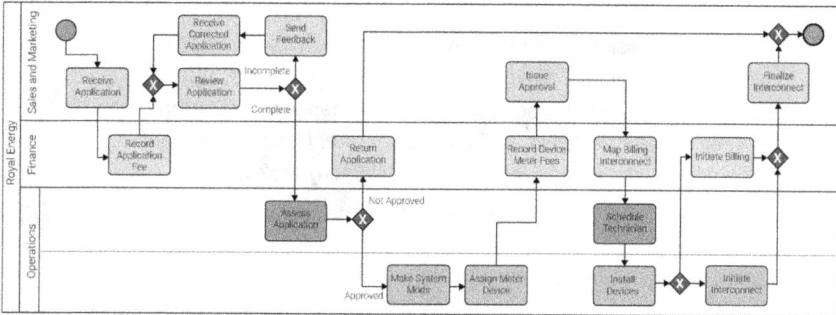

Notes

1. The team did finally create the process models with BPMN (business process modeling notation), an industry standard. See a tutorial at https://camunda.com/bpmn/.

2. To create the models, they could use any of the freely available as well as commercial modeling platforms such as http://demo.bpmn.io (designed and maintained by Camunda Services GmbH).

Reflection Questions

1. As the team starts by examining the process portfolio, the discussion moves from one process to the next until Claudia chimes in: "What about that interconnect process?" This is a proposal that Kevin seems to support adamantly. Do you agree with their choice? Why?

2. Later in the discussion, Abra offers this insightful commentary: "I think it is quite clear that no single individual will know the entire process. In fact, different departments can only tell you what they do and maybe what happens just before and right after." That seems to ring true—is it? Please elaborate. If this information is not readily forthcoming, what should the team at Royal Energy do?

3. Another issue that the group ponders is about "making pictures of processes" (Abra's words). Here, Abra considers the possibility of using the flowchart technique, but after much discussion, she realizes that "to construct a picture of the process, we will have to gather a lot of information" and "a flowchart won't be enough to accommodate this information." She appears to argue for a technique better suited for business processes. Your thoughts? Be specific!

Self-Assessment Questions

1. Can you explain the fundamental ideas underlying models of business processes?

2. Can you distinguish modeling approaches for business processes from others?

3. Can you build a process model to reflect information from stakeholders?

Readings

1. Object Management Group. 2016. "BPMN Quick Guide." Accessed January 31, 2024. https://www.bpmnquickguide.com/view-bpmn-quick-guide/.

2. Camunda. 2020. "Web-Based Tooling for BPMN, DMN and Forms." Accessed January 31, 2024. https://bpmn.io/.

3. Signavio. 2020. "BPMN Modeling Guidelines." Accessed January 31, 2024. https://www.modeling-guidelines.org/.

Key Terms

These definitions are interdependent. An iterative reading is recommended.

Atomic Task/Activity

An atomic task is one that cannot be broken down further. It is called atomic because it is indivisible. When executed, an atomic task either completes or fails; it cannot be done partially. Identifying atomic tasks is important for well-designed process models.

BPMN (Business Process Modeling Notation) Language

BPMN is a technology-agnostic standard language (introduced in 2006, managed by OMG) that provides a standard notation for *activities*, *swimlanes*, *gateways*, and more to represent the flow and control logic of a process.

Conceptual Model

The conceptual model (of a business process) is a high-level representation that provides an abstract view (of the process). It may not include detailed operational information. Instead, it focuses on the key elements such as *tasks* and *actors* and the overall flow of the process.

EPC (Event-Driven Process Chains) Language

EPC is a language (introduced during the 1990s in Germany and widely adopted in the SAP community) for business process modeling. It includes events, functions, organizational units, and control elements to represent the process flow.

Event

An event often represents the trigger that initiates a process or a path within the process. Examples include a customer placing an order or the arrival of a time. BPMN language also recognizes other kinds of events (that arrive in the middle of a process or signal the end of a process).

Gateway

A gateway in a *BPMN* process model is used to control *sequence flow*. One example is branching, or splitting a flow into multiple flows. An exclusive gateway will force a token to take only one path. An inclusive gateway will require a token to take all paths.

Lane

A lane is a partition within a *pool*. Lanes represent roles or departments within a larger entity such as a company. Lanes allow a model to group activities based on the responsible *actor*. Coordination across lanes is accomplished by *sequence flows*.

Message Flow

A message flow (represented by a dashed line and directed arrow in *BPMN*) shows messages as asynchronous communication between larger entities in a process. It is used to show communication between *pools*, not across *lanes* within a pool.

Modeling Language

A modeling language (for business processes) can be used for *process modeling*—that is, to create a graphical representation of a process. It provides syntax (e.g., standard notations) and grammatical rules to depict several process details.

Naming Conventions

- Name a *task* using a verb in the infinitive and an object (e.g., review draft, check invoice) without mentioning the actor.
- Name an *event* (whenever possible) using an object and a verb in the past tense (e.g., draft reviewed, invoice checked).
- Name an *exclusive gateway* with a question and label outgoing sequence flows with the conditions they are executed under.
- Name an *inclusive gateway* with a descriptive question and label outgoing sequence flows with the descriptors.
- Name a *pool* with a descriptor of the larger entity that describes the group of *lanes* it contains or a label that describes the part of the process succinctly.
- Name a *lane* within a *pool* with the *actor* (such as a role, individual, team, or automated system) responsible for the *activities* shown in the *lane*.

Petri Net Modeling Language

The Petri net modeling language is a mathematical modeling language (proposed in the 1960s) that has been widely used to visualize, simulate, and analyze business processes. A process model constructed with this language consists of places, transitions, and directed arcs.

Pool

A pool is a type of swimlane that represents major larger entities in a process such as a company, partner, or customer responsible for managing a self-contained set of activities. A pool is independent of other pools. Coordination across pools is accomplished by a *message flow*.

Process Instance

A process instance refers to a single execution of a process. For example, a process designed for handling orders will be executed differently for each order—that is, a separate process instance. Each process instance is unique in terms of data and, potentially, the *sequence flow*.

Sequence Flow

A sequence flow (represented by a solid line and directed arrow in *BPMN*) shows the order in which activities are performed in a *process*. A sequence flow cannot cross *pools* but can cross *lanes* (e.g., roles or departments within a company) to depict coordination among actors.

Swimlane

A swimlane is a visual element, shown with horizontal rectangles in *BPMN*, to organize activities based on the performer (individual, role, department, system). A swimlane may be a *pool* such as a company.

Token (or Case)

A token or a case, such as a specific order by a customer, represents the occasion to execute a *process instance*. A token moves along one of the many possible *sequence flows* in a process model. The movement of tokens illustrates the dynamic behavior of the process.

Scene 5
Analyzing a Process

We know the customers complain about that process
Let's try to locate what's wrong, what are the causes

Learning Objectives

- Describe and use some fundamental process performance metrics
- Describe and use techniques for qualitative analyses of process models
- Explain the need for process model simulations
- Interpret process model simulation results to identify problems

Characters

Regina Wood—VP, Eastern MA
Anik Malik—VP, New Hampshire
Samantha Bellman—VP, Western MA
David Ashley—VP, Rhode Island
Remy Garcia—IT, New Hampshire
Sophie Raymond—Operations, Eastern MA

Narrator: *We are outside the executive conference room in Cambridge, MA. It has been several weeks since the last meeting, when our rising stars discussed different ways of modeling a process and appeared to settle on one. Since then, they have identified an industry-standard modeling technique for business processes. Several groups have been working on creating models of some of the high-priority processes with this technique. One of these groups, represented by Sophie Raymond and Remy Garcia, was selected to present today because they have emerged as de facto experts. Over the last several weeks, Sophie and Remy have returned to what they had learned about business processes and added to that with sources on the web and elsewhere. They've helped other groups understand the ideas and have apparently produced some useful and detailed models and analyses for the interconnect process. Today, they are going to present some of these results. Sophie is outside the conference room door looking at her phone. Remy approaches.*

REMY:	Sophie!
SOPHIE:	Oh, thank goodness you made it.
REMY:	Oh, I've been here. I got the extra copies you needed. Mickey at the front desk is really nice and let me make some.
SOPHIE:	Mickey? I thought his name was Frank.
REMY:	Oh, I just call him Mickey. Doesn't he look like Mickey Mouse to you?
SOPHIE:	(*checking her phone*) Where in the world are they? We're about to start, and—
REMY:	Shh! Here they come.

(*Regina, Samantha, and Anik all approach*)

REGINA:	Good morning.
SAMANTHA:	Thank you for your patience. This morning's meeting went over. Shall we go in?

(*the group troops in and starts setting up*)

SOPHIE:	(*nervously*) So sorry; the door was locked, so we didn't have time to—
SAMANTHA:	It's not a problem.
SOPHIE:	Okay, good, great. Yes. (*covertly*) Remy, did you bring the . . . ?
REMY:	Oh man, I left it in my car.
SOPHIE:	What? I—
REMY:	Just kidding, I brought extras of everything. We're good.
SOPHIE:	(*mumbling*) Oh my gosh. (*out loud*) Okay, we are ready.
REGINA:	Great. Well . . . good morning, everyone! We know your work, but we don't know each other yet. I'm Regina Wood. This is Samantha Bellman and Anik Malik.
SOPHIE:	I'm Sophie Raymond.
REMY:	Remy Garcia.
SOPHIE:	I'm sorry, are we waiting for one more?
ANIK:	(*snickers*)
REGINA:	Ah, yes . . . well—
DAVID:	(*from the door*) Are you talking about me?

Narrator: *Everyone turns to look as David Ashley, another VP, saunters in.*

REGINA: Well, well, well . . .

DAVID: Regina! I knew you would wait for me. (*addressing the team*) David Ashley, VP of the Rhode Island division.

REMY: I'm Remy, and this is Sophie.

SOPHIE: Nice to meet you all.

(various murmurs and sounds of chairs being sat in)

DAVID: Shall we begin?

SOPHIE: So as you all know, we are just one of the many teams that have been building models of the 14 high-priority processes out of 236—

ANIK: To be specific, you are the best of these teams, are you not? And we had 8, not 14?

REGINA: Yes, that analysis was extended; I remember the report that came in. Abra and the others worked to locate 14 high-priority processes. That is what you're talking about, right?

SOPHIE: Yes. Now we're here to show you a model of our assigned process. So this is the process as it now stands. Remy?

REMY: (*presents a projection of the process on the wall*) This is what we've been looking for, a picture of the interconnect process. We are projecting the model directly from our software, so if there is something we need to change, we can modify it as needed and work with it in the moment.

REGINA: Impressive.

REMY: Thank you. What we found is that we need a lot of detail. I'll show you.

(Remy clicks a button and a dialogue box shows up)

ANIK: Oh . . .

SAMANTHA: We can't even read the chart—why is there so much detail?

ANIK: Yes, please explain all this.

SOPHIE: So first we need to know the expected volume of the process . . . such as seven hundred customer requests a day, and then there may be some variability. Like more requests on Mondays versus other days or requests on holidays versus other days of the week.

REMY: We also need to know who is responsible for each task and how many people are in each role. And that may be a person or a piece of software. We do that by placing different tasks in different lanes—and those names we see on the lanes are the departments responsible for those tasks.

SOPHIE: We also need to know the expected time for each task in the process. That we can do by entering the average time and standard deviation for each task.

REGINA: Well, I can't fault you for that. Obviously, we have to have those things to learn how the process is doing. Just seems like a lot of detail . . .

SOPHIE: Exactly. But how can we analyze the process if we don't have these details?

ANIK: Yes, but it's a lot. Frankly, it's overwhelming.

SOPHIE: I understand, but—

ANIK: Perhaps we don't need all of this information.

SAMANTHA: Yes, maybe we can figure out what we need and don't need.

SOPHIE: But if we want a complete—

REGINA: Prioritizing. That's what we need our next step to be.

DAVID: Y'all, the expert we hired is trying to speak. Go ahead, Sophie.

SOPHIE: Thank you, Mr. Ashley—

DAVID: Oh, you can just call me David. We're all colleagues here, aren't we?

SOPHIE: Alright . . . David . . . um, I was actually going to highlight an example Remy gave last week for why we need this detail. Remy, remember when you were talking about the parties you used to throw in Barcelona?

REMY: Oh, ah, yes. Well (*chuckles*), in Barcelona, there's a really strong nightlife scene; it's integral to the community. We often go out, but sometimes we plan a party at home. Let's say that we are modeling the process of how to plan a party. We'll have to figure out how long it takes to make food for the party. We have to know things, like how many dishes there will be and what they will be. *Bombas, paella, mató*—and each dish takes a different amount of time. Some start times will have to overlap with finish times, and so on. And of course, we have to know how many guests are coming, who is helping to make the food, who is bringing the wine, and if anybody has allergies.

DAVID: Well, now I'm starving!

REMY: (*laughs*) We also have to know the tools we have. Oven? Stovetop? Microwave? Depending on what is available, we either can or cannot make the food we have planned.

SOPHIE: My sister owns a restaurant, and the same detail applies there. To analyze a food-making process, we first have to know all the things on the menu. How many guests come in on a daily basis, even down to the dish everyone wants to order for Sunday brunch.

REGINA: I've got to admit, that makes a lot of sense.

SOPHIE: So even though the details seem overwhelming, they're crucial to obtaining a complete picture of the process. Without them, there will be problems later on.

DAVID: I agree with Sophie and Remy; we're already taking the time to do the work now. May as well do it right and save time later.

SAMANTHA: (*sigh*) Well, these are good points. But let's say we're going to analyze the process . . . what does that actually mean?

REMY: Well . . . we're still working on that.

ANIK: What do you mean? I thought you were ready to present today.

REMY: (*slightly annoyed*) We are. We've figured some things out, and there are some things we still need to—

ANIK: Well, let's see what you've figured out, then.

REMY: (*irritated*) Yes, that's what we—

SOPHIE: That's just what we planned. Let's just keep to the interconnect process, for time's sake. So one way to analyze a process is to look at a normal case. We can assume that we'll have seven hundred applications each month, and we can assume that out of those applications, 20 percent will be incomplete, needing revisions. The other 80 percent will be ready for assessment. Then of those, some 4 percent will have to be denied and returned. The other 96 percent will move forward in the process for interconnect, approval, and device installation. With these assumptions, and many more, we can analyze the process.

SAMANTHA: Hmm, okay . . .

SOPHIE: Then we can see—on average—how long it will take for the process to run. And we can see how long it will be for each new application that is well formulated and requires no corrections—

REMY: And we can see how many people we need on payroll to respond to this level of activity. So these are calculations we can do, and we can use this information for planning ahead. For example, if we know that over time, the volume will go up or those relative fractions will change—

ANIK: And before you say so, let me add this: on average!

SOPHIE: (*smiling*) Yes, these are calculations, so we use phrases like "the throughput rate" and "process capacity." For example, process capacity would be how many requests our process can handle given our current staffing. In our case, we found that our capacity is one thousand requests. Beyond that, some of our tasks will become overloaded.

ANIK: So we're utilizing 70 percent of process capacity now.

SOPHIE: Correct! And we can also calculate cycle time.

ANIK: What's that?

SOPHIE: That's the time it takes for a request to go from start to finish—through the process.

ANIK: On average.

SOPHIE: Yes!

REGINA: But things don't always go smoothly, do they? What happens if we have ups and downs during the day? I can imagine that mornings will be busier—and let's not forget that our employees won't always be doing things at top speed. Sometimes they may have to deal with solving problems that may make some tasks take longer than others.

Narrator: *Remy and Sophie exchange an excited look. Prior to the meeting, Sophie had wondered if they could introduce these complex ideas in a discussion. But the VPs are sharp and asking hard-hitting questions already. This is what the presenting team had hoped would happen, and they've brought their A-game! Sophie and Remy are prepared to respond to how the discussion is moving ahead.*

REMY: Yes, another way is to vary that normal case. That's the other way of doing process analysis. We can say we'll have seven hundred applications each month—on average—but some days, we'll have a couple hundred, and some other months, we may reach a peak of fifteen hundred. So we have to look at that variation. And *then* we can go a little deeper than that.

SAMANTHA: Oof, the detail again . . .

REGINA: Well, it's needed.

REMY: Even if we say there will be, for example, twenty applications a day, those applications won't be all spread evenly—say, like, two and a half applications an hour for an eight-hour day. The applications may arrive in bursts, so there may be none in the morning, a few in the afternoon, and then many more in the evening, etcetera. So that's the part we have to figure out . . . how to analyze a picture of something that changes all the time.

DAVID: So let me try to put this into layman's terms—

SAMANTHA: For whom? There aren't any laymen here.

DAVID: (*laughs*) For me, Sammy! I'm the late one here; I gotta make sure I'm following along.

(*chuckles from the group*)

DAVID: Okay, so, how I understand it, we can make an assumption about how many applications there will be in a month and then get a couple of rough percentages on whether the applications will be ready for assessment. Another way is to take variation into account . . . some months there will be more applications, some less.

ANIK: Seems like there are only two options, actually. The third is the same as the second; it just has more detail.

REMY: Well, that's a way of looking at it, yes . . .

SAMANTHA: Well, I'm all about hard work, but the first way seems easier to me. Not sure what this second way of looking at it gets us, with all these variations.

SOPHIE: True, but there are some things—

REGINA: I don't know. We came all this way . . . I think we should buckle down and do it in the more detailed way.

REMY: The trick is that all the data we discussed before will have to be in place before we can even try these different options for process analysis.

REGINA: What I've learned here is that the drama really is in the details. We've cracked open a big can of worms here, folks, and we've got to see it through.

ANIK: And we should specify that there are really *two* ways to analyze this. One is to look at a normal process and another is to vary that process. Exactly how that is done will have to be up to the teams.

REGINA: I'm still waiting to hear an answer to Samantha's question, though.

Narrator: *Remy and Sophie pause, unsure about what they missed. They look back at Samantha, who looks up from her notes.*

SAMANTHA: I think we understand the calculations . . . process capacity and something else you mentioned. We get that. The second way you mentioned—what is that? Is that not calculations? And more importantly—this is what I asked before—what does it get us that your first way doesn't?

REMY: (*rising to the challenge*) Simulations! That's the second way. So we run the process using software. Because it is a simulation, we can "pretend run" an entire day or a month in a few seconds. And when we do that, we can change things like how the requests are coming in with bursts and even change how the time taken to do a task can vary.

DAVID: That sounds fine, but . . . can we see these models running on screen?

REMY: Well, almost. They will generate statistics that we can see as hot spots on a map, and we can then analyze the numbers.

SAMANTHA: I'm still waiting to hear: What does it get us?

REMY: Oh, sorry—yes! Well, when we simulate it this way, we can see which parts of the process are getting overloaded, where the bottlenecks are. For example, if we see a large volume of requests coming in, and we see that the requests are all getting bunched up in some parts of the process with long lines waiting for the next task to be done, we have just analyzed what's called a bottleneck!

DAVID: I see the point here—

SAMANTHA: Yes, thank you for answering my question.

Narrator: *Samantha has digested the information from Remy and Sophie quickly. She needs no further explanation. She is convinced.*

SAMANTHA: (*abruptly*) Sophie, Remy, thank you for all this work you've done. And surprise, surprise, there's more to do now. Could you summarize for us what you presented here? Preferably on a single-page document.

DAVID: With bullet points! I love bullet points.

SAMANTHA: Yes, David, we all know about your love of bullet points . . .

REGINA: A summary will be necessary. We need to learn as much about how this works as we can, as we may be sharing the progress with the board next week.

SOPHIE: Of course! We'll get right on it.

ANIK: You also need to go back and look at the tools you're using and figure out exactly what kind of analysis they can do. Do you think you can do that?

REMY: (*bristling*) I have a—

SAMANTHA: Anik, stop being a grouch. Of course they can do it; we asked them to come in and present for a reason. They're the best of the best. Let me add this, Sophie, Remy: I came in here thinking you were going to tell us about things like lean and Six Sigma. I am somewhat familiar with that terminology. But what you showed here was different. It was more specific and to the point. I am impressed and look forward to where we can go next. That summary is important—and we want to make sure that our other teams are brought up to speed with these analytical approaches.

SOPHIE: Thank you!

REMY: Thank you, Samantha, for those kind words!

ANIK: Wait . . . can't we still use those things like lean and Six Sigma for process analysis? I'm sure we don't want to just throw that away, right?

SAMANTHA: Anik, nobody is throwing anything away. That Six Sigma analysis is a statistical technique to analyze process outcomes. What we have seen from Sophie and Remy is about the internals of the process.

ANIK: (*protesting*) But we still need to look at the outcomes, right?

SAMANTHA: (*reassuring*) We will, Anik, we will . . .

REGINA: Well, I think we're done here. Good job, you two!

(The meeting wraps up and everyone goes their separate ways. Later, in the parking garage . . .)

REMY: Anik . . . Anik! You made me look so incompetent. That was my first meeting with the other VPs!

ANIK: I know, I know. I didn't want them to think I got you the job or something.

REMY: Anik, you *did* get me the job!

ANIK: I only made the referral. You were the best person for the job. I will try to be nicer.

REMY: I'll forgive you only if you and Joe invite me over for dinner.

ANIK: (*laughs*) I think we can manage that!

To be continued . . .

Appendix 5A. Process Analysis Metrics

Metric	Value for the Interconnect Process
Cycle Time	The process takes, on average, 15 days from the start (receipt of an application) to the end (resolution of that application)—that is, the *cycle time* for the interconnect process is 15 days.
Process Capacity	At the current staffing levels, a maximum of 1,000 applications can move from start to end every month—that is, the *process capacity* of the interconnect process at the current staffing levels is 1,000 applications for each month.
Capacity Utilization	If the company anticipates, on average, 700 applications a month, they can process these applications without feeling overburdened—that is, the *capacity utilization* of the interconnect process, at current volumes, is 70%. (Note the process capacity of 1,000 applications above.)
Resource Utilization	Some tasks (done by the staff in Operations) take more time, other tasks (done by the staff in Finance) take a little less time, and yet others (done by the staff in Sales and Marketing) take even less time. As an example, the "Financial Analyst" is responsible for "Record Application Fees" and "Record Device Meter Fees" (see the process diagram in appendix 4A). These two tasks require 5 minutes and 2 minutes, respectively. So the analyst spends 7 minutes on each application. For the month (20 days, 7 hours a day, 60 minutes each hour), she can handle 1,200 applications—that is, the *resource utilization* for this role, at current volumes (700 applications), is 58.33%. Different roles will have different levels of resource utilization. (Note that resource utilization is not the same as capacity utilization.)
Flow Efficiency	Although it takes an average of 15 days for an application to go through (cycle time), the customers see "value received" for only 3 days, when they submit the application, get a status update, and receive the final resolution. This means the process has a flow efficiency rating of 20%.

Notes

1. The table shows fundamental process analysis metrics for the interconnect process. It describes each metric and illustrates it for the interconnect process.

2. Detailed computations for the metrics are not shown here. The illustrative examples provide an intuitive understanding.

Appendix 5B. Process Simulation Results

Indicators	Value for the Interconnect Process
Wait Times	The simulation results show that the tasks "Make System Mods" and "Install Devices" have the longest wait times (2 days and 4 days, respectively). This result suggests that, prima facie, these tasks may be bottlenecks. These tasks may need careful examination to improve process capacity.
Durations	The simulation results show that the tasks "Assess Application" and "Schedule Technician" take the most time (2.3 and 4.5 days, respectively). These tasks may need careful examination to improve process cycle time.
Costs	The simulation results show that the tasks "Return Application" and "Initiate Billing" are the most expensive ($150 and $220, respectively). These two tasks may need careful examination to reduce process costs.

Notes

1. Process analysis metrics can only show averages or aggregates. However, there can be variations (e.g., a large number of applications may arrive together). To understand how the process will behave, we can simulate a process.

2. The team used an online simulator: https://bimp.cs.ut.ee/simulator. The tool allows one to upload a standard BPMN process model. The team uploaded process models such as the one for the interconnect process (see appendix 4A). The tool checks the model for internal consistency and allows the team to enter parameters to run a simulation.

3. The team entered the following parameters:

   a. a scenario specification—for example, seven hundred applications, with one application arriving every fifteen minutes and a standard deviation of seven minutes

   b. the number of individuals in reach role—for example, one financial analyst, two clerks, and so on along with an hourly rate for each role

   c. a time for each task—for example, five minutes for the task "record application fees" with a standard deviation of two minutes

   d. the percentages for any splits—for example, after the task "install device," 30 percent of the applications move to the task "initiate billing," and 70 percent of the applications move to the task "initiate interconnect"

4. When a simulation is run, the results can vary. For example, applications may not arrive at regular intervals (every fifteen minutes). Sometimes, the time between two application arrivals may be short (five minutes); at other times, it may be very long (twenty minutes). A task such as "record application fees" for an application may need five minutes or less (three minutes) or longer (say, nine minutes).

5. For the interconnect process, most parameters did not have large standard deviations, and therefore, the results did not vary greatly for each simulation run. The illustrative examples provide an intuitive understanding of how the simulation results can be used for process improvement.

Reflection Questions

1. As a part of their discussion, the group ponders one specific issue a few times in their quest for process analysis. It is captured by Anik thus: "Perhaps we don't need all of this information." Information such as the volume of incoming requests, staffing levels, fractions for different paths, times for different tasks, and perhaps much more. Does the team at Royal Energy really need to do all of this tedious work? What do you think?

2. Another important discussion appears to focus on the phrase "on average." What's the point here? Why should the team be worried about this as part of their process analysis efforts? Should they or should they not consider things "on average"? Why?

3. During the scenario, Remy and Sofie use a number of different words and phrases that point to key process performance metrics: "cycle time," "process capacity," "flow efficiency," and others. Develop a table that shows (a) a definition for the metric (you may search for this) and (b) how Royal Energy can use this metric.

Self-Assessment Questions

1. Can you describe and use fundamental process performance metrics?
2. Can you describe and use techniques for qualitative analyses of process models?
3. Can you explain the need for process model simulations?
4. Can you interpret process model simulation results to identify problems?

Readings

1. Dumas M., M. La Rosa, J. Mendling, and H. A. Reijers. 2018. "Quantitative Process Analysis." In *Fundamentals of Business Process Management*, 255–96. Heidelberg: Springer Berlin.

2. BIMP. 2020. "Business Process Simulator with BPMN." Accessed January 31, 2024. https://bimp.cs.ut.ee/simulator/.

Key Terms

Business-Value-Added Tasks
Business-value-added tasks may be necessary for the *process* for reasons such as regulatory requirements or other factors. These may be targets for cost, time, or effort reduction.

Capacity Utilization
Capacity utilization refers to the degree to which a process is utilizing its potential *capacity*. It is expressed as a percentage. If at or near 100 percent, demand increase can cause delays; if there is unused *capacity*, it could mean resource underutilization.

CTQ (Critical to Quality)

CTQ are characteristics of a product/service that are important to the customer. They must meet or exceed the customer's expectations for the product/service to be considered high quality. CTQ is important for process improvement efforts such as *Six Sigma*.

Cycle Time

Cycle time is calculated as the time needed for process execution for one *process instance*. For example, if *run time* is 54 minutes and twelve *process instances* move concurrently through the process, then it is as if it takes 4.5 minutes for each process instance. That is the cycle time.

DMAIC (Define, Measure, Analyze, Improve, and Control)

DMAIC is a data-driven quality strategy for process improvement. It is a core component of the *Six Sigma* approach but can be implemented as a stand-alone quality-improvement approach.

Interarrival Times (for Process Instances)

The interarrival time is the time between successive arrivals of *process instances*. For example, if a business (open 9 a.m.–4 p.m.) receives one hundred customers, then the interarrival time is 4.2 minutes (may be specified as normal/exponential distribution with appropriate parameters).

Lean Process Improvement

Lean process improvement is an approach (originating at Toyota in the mid-twentieth century) to reduce waste by eliminating *nonvalue-added tasks* such as wasted resources, *wait times*, unnecessary movement or transport, overproduction, defects, and excess inventory.

Nonvalue-Added Tasks

Nonvalue-added tasks do not directly contribute to the creation of customer value. They do not add value from the customer's perspective and may include activities like transport, storage, and inspection. These can be targets for elimination.

Path Probability

Path probability refers to the likelihood of a path a *process instance* will take when it arrives at an exclusive *gateway*. For example, in a loan-approval process, paths for loans < $50,000 and > $50,000 may be different, where path probabilities may be stated as 75 percent and 25 percent, respectively.

Process Capacity

Process capacity refers to the limit on the *throughput*—the maximum number of *process instances* a process can handle over a given period of time. It is influenced by the resources available and the design of the process itself.

Process Simulation

Process simulation is a technique that mimics process execution based on estimated parameters such as *task times*, *path probabilities*, *actors*, *interarrival times for process instances*, and other details. A software tool then generates *process instances*, mimics the *sequence flow* of each, and tracks various performance measures.

Resource Utilization

Resource utilization refers to the degree to which resources (machinery/equipment/employees) involved in a process are being used. It is measured as a percentage—for example, if an employee works six hours in an eight-hour shift, their resource utilization would be 75 percent.

Run Time

Run time refers to the time it takes to execute a *process* from start to finish. It begins with the start *event* and includes the time to perform all *tasks* in the process (plus any *wait time* between tasks) before reaching the end of the process.

SIPOC (Suppliers, Inputs, Process, Outputs, and Customers)

SIPOC is a type of process mapping tool for process improvement methodologies like *Six Sigma*. A SIPOC diagram provides a big-picture view by showing the key elements and their relationships.

Six Sigma

Six Sigma is a data-driven methodology for eliminating defects and improving quality in high-volume processes by measuring defects and then calibrating the process to minimize variations and reduce defects.

Task Time

Task time refers to the time it takes to complete a *task* or *activity* within a *process*. Standard task times can be used to compute statistics such as *run time* and others or may be specified using a distribution (e.g., normal distribution) for *process simulation* to estimate performance.

Throughput

Throughput refers to the number of *process instances* that are completed in a given period of time. It is a measure of the productive capacity. For example, if *cycle time* is 4.5 minutes, then the throughput of the process that runs for nine hours would be 120 *process instances*.

Value-Added Tasks

Value-added tasks directly contribute to the creation of a product or service in a way that the customer is willing to pay for. In other words, these *tasks* add value from the customer's perspective.

Wait Time

Wait time is the period during which a *process instance* is not actively being worked on—that is, it is waiting for some *event* to occur or a precursor concurrent *task* to be completed or some resource to become available. Excessive or unnecessary wait time can reduce efficiency.

Scene 6
Redesigning a Process

How do we do this? Consider analysis results, use the heuristics
There will always be trade-offs, but it will be a potent mix

Learning Objectives

- Decide on the appropriateness of process improvement strategies based on context
- Apply different strategies and heuristics for process improvement
- Distinguish between redesign and reengineering approaches
- Locate trade-offs inherent in process improvement

Characters

Regina Wood—VP, Eastern MA
Anik Malik—VP, New Hampshire
Samantha Bellman—VP, Western MA
David Ashley—VP, Rhode Island
Remy Garcia—IT, New Hampshire
Sophie Raymond—Operations, Eastern MA

Narrator: *It's been a few weeks since the meeting in our last scenario ended, and the teams have been analyzing the processes. But now they have to think of how to improve the processes based on the results. That's what we have on the agenda. We are in the boardroom. The four VPs are seated around the table. Sophie is happy and confident today, chatting with David and the other VPs. Remy is running late, but nobody seems to be worried.*

DAVID: . . . and then I swung too hard and lost my club! I swear, I made a hole in one with that club flying through the air by mistake. Never hit the golf ball one time.

SAMANTHA: I can't believe you hadn't ever played golf before! Isn't that just sort of standard with being a fancy-pants with a fancy-pants job?

REMY: *(rushes in)* So, so, so sorry I'm late.

REGINA: Don't worry about it. Take a breather and we'll all get started.

REMY:	Thank you, Regina. Sophie, can we go over some of these things for a second?
SOPHIE:	Sure thing. (*privately*) Remy, having a bad morning?
REMY:	Just running late . . . let's just get this over with.
SOPHIE:	Okay . . .
REMY:	I'm okay, really. Let's do this!
Narrator:	*Sophie and Remy pull their work together, whisper to each other for some last-minute changes, and confirm their plan for the meeting. Samantha kicks off the discussion.*
SAMANTHA:	Welcome back, everybody. Let's get right to it. Remy? Sophie? What do you have for us?
SOPHIE:	As we all remember from our last meeting, our goal was to find a way to analyze the process. This is so that we can learn exactly how it works, and what does or doesn't work, so that we can fix whatever needs fixing. Well, we have the analysis results! We know how the process works and where the problems may be. Now we get to move on to the next step: redesign.
DAVID:	Right, the fun part! So I hear different words—there's *improvement*, there's *redesign*, and . . .
SOPHIE:	Actually, I think all of those have to do with how extensive the change is—we will get to that.
REMY:	Yes, now we have to decide how we want to do this fun part. We have options: one is to leave the basic process structure unchanged—so, the tasks and such—and make some things faster or add more people.
DAVID:	So this would be incremental improvement—small changes!
SOPHIE:	True! For example, a restaurant may find some way to speed up chopping vegetables, add an extra oven, or add more staff. All of these will be small changes—or, as you said, incremental improvements. Sometimes, these small changes can make a big difference in how the process works.
REMY:	Option two would be much more radical. We can change the very basis of how the process is done, essentially making an entirely new process. I recently came across an example. Have any of you been to Spyce, the new restaurant downtown?
REGINA:	Oh, my son was talking about that. The one with the robotic kitchen. It's not so new anymore, is it?

REMY: Well, I had never heard of it, so it was new to me! They designed their entire structure around the idea of robots cooking the food. It allows them to sell quality meals at a low price because the machines are doing all of the cooking. And the food really is delicious.

ANIK: It's such a huge step to decide not to hire line cooks . . .

REMY: Well, someone has to prep all the ingredients before the cooking, but other than that, I bet they've found it's very efficient.

SOPHIE: So that's an example of the complete redesign of a process—

SAMANTHA: I wish we could send in some robots to solve our problems for us.

SOPHIE: Good point; that's actually what I wanted to bring up. It's very tempting to try to redesign the process or just make small improvements by throwing technology at a problem. But we all know that it's more complicated than that. Think of this interconnect process we've been working with. We spoke with different stakeholders, and they all just wanted us to add technology and make the problems go away. We had to explain that it doesn't work like that.

REMY: I would say that Sophie was pretty forceful about her opinion on this . . .

SOPHIE: I was firm, yes. But it's important for everyone involved to realize that these processes need careful work, not just a high-tech Band-Aid.

SAMANTHA: This reminds me of when my sister gave me a wonderful Christmas present: a state-of-the-art electric drawing and sketching tablet. Any paintbrush or art tool you can think of could be mimicked with a screen and a stylus. But I spent more time trying to figure out how to use the darn thing than I did painting. In the end, nothing comes close to my trusty Prismacolors.

DAVID: You do art?

SAMANTHA: (*chuckling*) I "do" art.

ANIK: But adding technology is important. Coloring pencils can only get us so far.

SAMANTHA: (*dryly*) I resent that.

ANIK: Okay, okay—remember faxing? Remember when sending a fax was a thing? We all did it . . . that dial-up sound, the waiting . . . but it was the most efficient way to get a document across the country. Sometimes even across the globe.

REMY: I'm sorry, I . . . I have never actually used a fax machine . . .

ANIK:	Ah . . . you're too young to remember fax machines. I'm officially an old man.
REGINA:	Remy, a fax machine would scan whatever document you chose and send it to another fax machine. If you had the fax number, you could send that machine any printed document you liked.
ANIK:	So think of the steps. Print the document, hand it to a secretary who will schedule time, then fax the document, maybe call them to make sure that they received the document, and so on. That's a miniprocess, isn't it?
REMY:	Oh, well, there's an app for that now.
Narrator:	*The group laughs.*
REMY:	What's so funny?
REGINA:	Well, you've actually hit the nail right on the head there.
ANIK:	Yes, now we can just pull out the phone, scan the document, and it generates a PDF that we can email. The whole faxing process is simply gone.
REGINA:	Rather like your *radical* redesign with those robots cooking your meal.
SAMANTHA:	So if we're *not* just going to throw technology at it, how should this be done?
SOPHIE:	That is exactly what we wanted to discuss, and we have some recommendations.
REMY:	The first is that we must listen to the experts who know about the process. To be clear, these are the people who have worked on the process and with the customers inside this process.
SOPHIE:	But there can be a difficulty. The experts may simply say "This is how we've always done things." So we have to listen to the experts but not let them run the process redesign.
DAVID:	I see . . . listen to them, but push them.
SOPHIE:	Correct!
REMY:	Another recommendation is to look at the analysis results. This seems like a no-brainer, but we didn't want to leave it out.
SOPHIE:	The results will tell us which steps in the process are too slow—like where the customers wait the longest and where the bottlenecks are.
DAVID:	Well, the analysis will tell us what the problems are but not the possible solutions . . . isn't that right?

SOPHIE: Yes, that's true. For example, if the process analysis reveals a bottle-neck at one step in the process, that's all it can do: reveal. But *we* can respond in a few different ways. We used the analysis to see where or when the bottleneck happens—let's say when the customer first calls in and they have to wait for a long time. Now we have to decide how to make the process better: redesign or improve.

DAVID: I see. We can hire more people to answer the phone . . . we can give the phone people a script to minimize chitchat. That's two solutions . . .

SOPHIE: We can write a FAQ on our website and encourage customers to go there first. That would reduce the number of incoming calls.

DAVID: Three options already . . .

REMY: Let's actually keep brainstorming this; this is good practice. We will be doing more of this in the future.

REGINA: Well, here's another. Let's take that first step when the call comes in and it is answered. We can break it up so that when someone takes the call, they can check if it's a simple question that can be answered quickly. And if it isn't, they can send the customer to an expert who is more prepared for a longer call. That way we can protect the experts so they can deal with only the difficult calls.

SAMANTHA: Or we can add texting capabilities for customer service. Some customers are like me . . . I personally hate talking on the phone. I'd much rather text or email. This way, our customer service team can multitask; they can help multiple customers at the same time if they're on text or chat instead of staying on a dedicated phone call.

REMY: Five possibilities. Keep them coming!

ANIK: We can add the texting capability, yes . . . *and* we can also have some artificial intelligence program automatically respond to the text for at least the first part of the conversation.

REMY: Yes, "automagically."

ANIK: . . .

REMY: It's a joke, um . . .

ANIK: (*amused*) Yes . . .

REMY: Anyway, so good work! We thought of six possibilities, and there may be even more.

REGINA: Is there some way to identify these possibilities? A systematic way to do it?

SAMANTHA: Took the question right out of my mouth, Regina. There must be a method somewhere—a guideline, perhaps?

SOPHIE: We don't know of a common method to identify these possibilities yet, but we think there must be some knowledge out there already. And that's our third recommendation: doing some research to find some general lessons we can draw on.

ANIK: Wait, wait. So "general lessons" . . . that means these are not *rules*, right? I mean, they can't be.

REMY: Right, these can't be rules that we follow blindly. If coming up with innovative changes to a process could be done by following rules, we'd have no jobs, isn't that right?

SOPHIE: Yes, these improvements and changes are quite specific—a lot depends on what the process analysis tells us, what we can do in our organization, and how much investment we're willing to make. Not forgetting the risks we're willing to take. We cannot come up with these improvement options unless we really look at the process very carefully.

REMY: I assume that's what Samantha meant by *guideline*—these are not rules that we can follow blindly.

ANIK: (*smiling*) A *guideline*, by definition, is some information or advice about how to do something—something that still requires judgment, so it cannot be applied in a rote manner. Thank you, Google! So yes. You're exactly right.

REGINA: Anyway, Remy, I agree. The best we can do is to see if we can find some guidelines or heuristics. And more importantly, they can't and won't give us the complete solution in any situation. Your team will need to do a lot of creative work.

SOPHIE: Okay, so let's say that there's a guideline or heuristic that says, "Look for parallel steps." So within the interconnect process, we may have to look at tasks like assigning a recharge plan, developing credit assessment, and scheduling technicians for installing. The first two tasks are done by the same group; the third is done by a different group. If we can find a way to have different employees do the first two tasks—maybe they can be done in parallel instead of in sequence.

DAVID: That's a mouthful! But I see where you're heading.

REMY: So that's about parallel tasks, and there are several such rules. We can consider each of these, but we have to see how they work for the process we are working with.

ANIK:	So the point is that there is work to do, got it. I understand this will require an effort. Is there anything else we should know about?
SOPHIE:	Yes, there is, actually—and this is an important point. We have to realize that in order to improve the process, we need to make some trade-offs. We have to realize that any improvement we make will promise an upside, but it may have a built-in potential downside.
SAMANTHA:	Wait, aren't we supposed to be improving the process? And that means we're making it better, right? That means the newly rede-signed process will be better—isn't that the idea?
SOPHIE:	Yes, but something as simple as hiring more people to make some steps faster may be technically successful and make the process work faster, but costs will go up as well. Or . . . let's say we implement the texting capability we discussed. That's going to increase flexibility, right? But some customers may not like it, and they may feel that the quality of our service is lower.
REMY:	Plus the extra cost of adding this technology. So the idea of the trade-off is that we will gain some things and we'll have to give up on some things.
ANIK:	But what about something like adding smart AI software that will make some tasks run much more quickly? That will help make the process faster, and the software will replace some of our labor costs, so those will go down as well, right?
SOPHIE:	(*thoughtful*) Hmm . . . it depends on the process, but let me suggest this: Won't some of the customers feel like there is less of a personal touch because you just automated this step?
ANIK:	I see your point . . . we have to acknowledge the give-and-take. Agreed.
REGINA:	Exactly.
DAVID:	You know, this sounds random, but I swear it's relevant . . . have any of you visited Amazon Go? That new supermarket?
SAMANTHA:	You mean Whole Foods?
DAVID:	I know Amazon acquired Whole Foods, but that's not what I mean. Amazon Go is a completely redesigned shopping process that has eliminated the checkout line.
	(*the following lines can be said simultaneously*)
ANIK:	No checkout lines?
SAMANTHA:	What?
REGINA:	How is that possible?

SOPHIE:	How do they pay for stuff?
REMY:	No way . . .
DAVID:	Yeah. Apparently they have these sensors that monitor which items you take off the shelves, and if you put it back, it's taken off your bill! Once you leave the store, the card you have on file through Amazon is automatically . . . or "automagically" . . . charged for the items you've chosen. Plus a receipt is sent directly to your phone.
SAMANTHA:	How have I not heard of this?
DAVID:	Amazon has eighteen stores already in major American cities, six in NYC alone.
SOPHIE:	Wow, talk about radical redesign!
DAVID:	Yeah, it's great for customers and for Amazon . . . not so great for other stores. This idea seems to be really taking off. Can you guess how long someone spends in a store like that?
SOPHIE:	I mean, sounds like five seconds!
DAVID:	(*laughing*) That's the right idea! Very little time, because you can walk in, pick up what you want, and just walk out. If you want something else, just come back and pick it up, then leave.
SAMANTHA:	Well, if we can do something as cool as that, our competitors won't even know what hit them!
REGINA:	So ultimately, we have to come up with innovative ways to redesign the process and solve problems.
SOPHIE:	Yes. We need to move away from the idea that we're improving the process and focus on redesign.
REGINA:	I think we've covered a lot in this meeting, though I feel that our work isn't quite certain.
SOPHIE:	As long as we understand that this is not a mechanical process and that we need to encourage creativity and innovation, we're halfway there!
SAMANTHA:	You've certainly given us a lot to think about.
ANIK:	I am a little worried about this. You've told us that this will require creativity, and more importantly, any change in the process will still need a trade-off. That means this is going to be quite open-ended.
DAVID:	I guess our people will have to take some risks to suggest their ideas. But can we shelf this thought for next time? I think this is a good place to wrap up our discussion today.

To be continued . . .

Appendix 6A. Improving the Interconnect Process

#	Improvement	Description	Based On
1	Hire more staff	Hire more staff in the Operations Department.	<u>Process analysis metrics</u> The resource utilization for staff in Operations was very high (~96%). <u>Process simulation results</u> The tasks "Make System Mods" and "Install Devices" done by staff in Operations have long wait times (2 days and 4 days, respectively).
2	Exploit concurrency	Redesign the process so the tasks "Schedule Technician" and "Map Billing Interconnect" are not done in sequence.	<u>Process simulation results</u> The "Schedule Technician" task takes a long time (4.5 days). <u>Process improvement heuristic</u> Place tasks that can be done concurrently (by different roles) on parallel paths in the process.
3	Separate special cases	Redesign the process to deal with problematic applications with a new path that contains two new tasks: "Send Feedback" and "Receive Corrected Application."	<u>Feedback from stakeholders</u> Stakeholders suggest that about 15% of the applications can benefit from targeted feedback so they can reenter in a revised form. <u>Process improvement heuristic</u> Cases that require special handling can be addressed via separate paths with an or-split in the process.
4	Split tasks	The "Return Application" task can be split into "Return Application" (moved to Sales and Marketing) and "Close Account" (retained in Finance).	<u>Process simulation results</u> The "Return Application" task, performed by staff in the Finance Department, is an expensive task. A part of the work can be broken off and moved to the Sales and Marketing Department. <u>Process improvement heuristic</u> For tasks that are complex, consider breaking them into multiple tasks.

Appendix 6B. Exploring Trade-Offs

#	Improvement	Trade-Off
1	Hire more staff	Hiring more staff is likely to (a) improve process performance by reducing cycle time but (b) increase costs
2	Exploit concurrency	Redesigning to exploit concurrency will (a) improve process performance by reducing cycle time, but (b) staff from the Finance and Operations Departments will need to be retrained and will have to learn to coordinate
3	Separate special cases	Redesigning to separate special cases is likely to (a) improve responsiveness to customers, but (b) staff in Sales and Marketing will need to take on the additional task of giving feedback, which may increase cycle time
4	Split tasks	Redesigning by splitting tasks is likely to (a) reduce cost by moving part of the work to lower-cost staff, but (b) staff in Sales and Marketing will need to be trained to do the new task

Appendix 6C. The Improved Interconnect Process

The improved process appears below. The numbers correspond to the improvements.

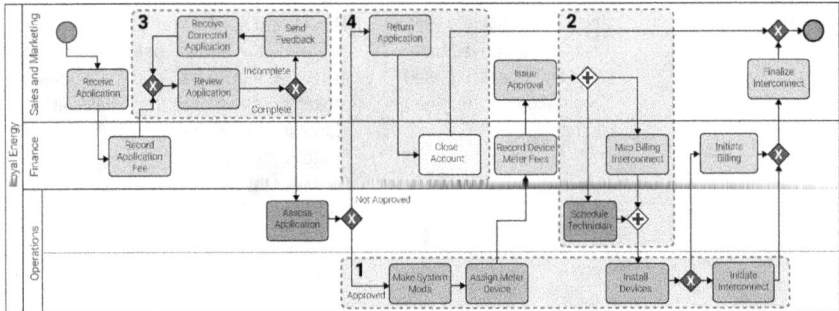

Reflection Questions

1. As this scenario begins, we hear Sophie say, "It's very tempting to try to redesign the process or just make small improvements by throwing technology at a problem. But we all know that it's more complicated than that." What does that mean for the interconnect process (or other processes) at Royal Energy? Be specific.

2. During the discussion, we hear about incremental improvement and radical redesign as two approaches to refining the process. Which strategy should the team at Royal Energy pursue? Why?

3. Can the team just listen to the experts at Royal Energy to improve the process? Should the team read and interpret the analysis results to improve the process? Should they do both? Why? Why not? And how should the team do this? Provide some examples.

4. When considering an opportunity for process improvement, the participants in the scenario kept coming up with possibilities. Remy declares, "Five possibilities. Keep them coming!" Finally, Regina asks, "Is there some way to identify these possibilities? A systematic way to do it?" So, is there? Provide the team at Royal Energy with a specific answer.

5. As the scenario progressed, the participants pondered another important point. It is Sophie who initially voices this: "We have to realize that any improvement we make will promise an upside, but it may have a built-in potential downside." This was later acknowledged by Anik as "the give-and-take." Should the team at Royal Energy be prepared for this? Explain how this may come about using the interconnect process as an example.

Self-Assessment Questions

1. Can you assess different process improvement strategies based on context?

2. Can you apply different strategies and heuristics for process improvement?

3. Can you distinguish between process redesign and reengineering approaches?

4. Can you locate trade-offs inherent in process improvement?

Readings

1. Reijers, H. A., and S. L. Mansar. 2005. "Best Practices in Business Process Redesign: An Overview and Qualitative Evaluation of Successful Redesign Heuristics." *Omega* 33 (4): 283–306.

2. Jan Vethman, A. 2018. "Shifting Trade-Offs in Business Processes." LinkedIn, January 23, 2018. https://www.linkedin.com/pulse/shifting-trade-offs-business -processes-ard-jan-vethman/.

3. Dumas, M., M. La Rosa, J. Mendling, and H. A. Reijers. 2018. "Quantitative Process Analysis." In *Fundamentals of Business Process Management*, 501–7. Heidelberg: Springer Berlin.

Key Terms

These heuristics are based on the discussion in Dumas et al. (2018), appendix A.

Customer-Focused Improvement Heuristics

Customer-focused improvement heuristics describe heuristics / rules of thumb that are aimed at improving the role of or interaction of customers, such as relocating control to the customer, reducing unnecessary customer contact, and integration (e.g., between your sales process and the customer's purchase process).

External Parties–Based Improvement Heuristics

External parties–based improvement heuristics focus on adding reliance on or improving communication and collaboration with external parties, such as the use of external trusted parties for some tasks, outsourcing a part of the business process to an external party, or providing a standardized interface for customers and other parties to engage in information exchange.

Information-Based Improvement Heuristics

Information-based improvement heuristics focus on the information consumed and generated in a *process*, such as adding controls to check accuracy before releasing information generated and buffering information so it can be accessed as needed for performing a *task*.

Organizational Resource Allocation–Based Improvement Heuristics

Organizational resource allocation–based improvement heuristics focus on the number of *actors* available to perform different *tasks*, such as assigning extra individuals to a role, providing training to employees, and empowering employees with greater discretion and decision-making authority.

Organizational Structure–Based Improvement Heuristics

Organizational structure–based improvement heuristics focus on improving how *actors* and other resources are allocated to different *tasks* in the *process*, such as minimizing how many different actors are assigned to tasks, keeping the assignment of actors to tasks flexible, avoiding shared responsibility for a task among multiple actors, considering mechanisms such as cross-functional teams and case managers, and hiring more employees in a role.

Process Behavior Improvement Heuristics

Process behavior improvement heuristics focus on improving the *sequence flow* of *tasks* (activities) in a process, such as resequencing tasks to leverage synergies, placing tasks in parallel sequence flows, ordering knock-out tasks in increasing level of effort, and separating exceptional cases from the normal process.

Process Improvement Heuristics

Process improvement heuristics are rules of thumb / lessons learned from prior process improvement projects that can provide pointers for deriving an improved process from an existing one in a new project. Although successful prior application is no guarantee for similar results in a new project, the heuristics provide a good starting point.

Task-Focused Improvement Heuristics

Task-focused improvement heuristics focus on improving the *tasks* (activities) in a process, such as introducing tasks that are focused on different *case types*, eliminating unnecessary tasks, executing tasks with a focus on each process instance, using triage for low-demand process instances, and considering activity composition.

Technology-Based Improvement Heuristics

Technology-based improvement heuristics focus on the use of technology for improving a *process*, such as automating a *task* or activity or making technology an integral part of the process such that a task or process behavior becomes superfluous.

A C T I I I

Considering

Technological Solutions

Scene 7
Deploying Processes

So you have redesigned some processes—great, let's deploy
Is there a technology platform I need to employ?

Learning Objectives

- Describe the need for specialized software for deploying business processes
- Explain the need for different components in this specialized software
- Describe how these components can help address commonly occurring problems

Characters

Regina Wood—VP, Eastern MA
Samantha Bellman—VP, Western MA
David Ashley—VP, Rhode Island
Claudia Narez—Environmental, Eastern MA
Abra McGregor—Finance, Western MA
Shalini Sharma—Sonita, a BPMS Vendor

Narrator: *There has been considerable activity since the last meeting. Remy and Sophie have been fired up, and they communicated this enthusiasm to the other teams. Each team worked on improving some of the key processes they were working with. Recall that these were the high-priority processes from the portfolio work that the organization did several weeks ago. This organizational effort, as well as the personal stories of our* *players, has evolved considerably since our story first began. Today, we're back at the executive conference room for an early, light breakfast before the meeting begins. Everyone's thinking, "What now? We have to figure out what to do with all these redesigned processes. They can't stay just as models on paper, can they?" The mood isn't jovial this time; everyone's sort of in a bad mood and a little tense. The winds of change are blowing. Right now, though, they're finishing up eating and ready to start . . . all except Anik, who is absent.*

REGINA: (*slightly irritated, with a bit of a headache that is turning into a lot of a headache*) If you all don't mind, I'd like to get this meeting going.

CLAUDIA: Of course.

REGINA: And I probably shouldn't be the one asking this . . . but what exactly are we trying to do here today?

DAVID: Regina, you stole my line.

SAMANTHA: The last time we met, we were trying to get a better picture of what the process looks like. Now we're trying to move past pictures and models—I think we are supposed to figure out how to start a rollout of processes into Royal Energy.

DAVID: So we have the firewood, the kindling, and the match . . . all we have to do is strike it?

CLAUDIA: No match yet, but that's where Shalini comes in.

Narrator: *Shalini has been sitting next to Abra as they both work through emails. The young vendor rep has acknowledged people as they've walked in but is nonetheless engrossed in her work. Abra doesn't seem to mind that Shalini is catching up on premeeting correspondence, but Claudia is nervous. Unlike most vendor representatives, Shalini actually comes with a reputation. A tech genius, she's very good at her job, and many people follow her posts about process management on her blog and social media. However, Shalini's blunt, brisk energy can be off-putting to some people. Privately feeling a bit protective, Claudia hopes the young woman makes a good impression.*

SHALINI: Shalini Sharma, I'm the technical representative from Sonita.

ABRA: Ah, so we *are* trying to "throw technology at it" . . .

CLAUDIA: (*surprised at Abra's sudden attitude*) Uh, this is a little different. We are not throwing technology at a specific process. I believe we are thinking of investing in a new technological platform for the company that will help us roll out several processes.

ABRA: So at some point, I think we will talk about why we need outside technology, right? Jamie and I have been—

REGINA: Because it's easier to get something that's ready and battle-tested than to start from scratch.

DAVID: The work you've done and continue to do is appreciated, Abra. And we haven't decided yet that we want to go with outside tech. We're still . . . looking at our options.

CLAUDIA: But we *are* ready to move on from models, analytical results, and proposals for redesigned processes to actual deployment. Now, before we listen to what Shalini has to say, let's bring her up to speed. We should tell her what we have done so far and maybe talk about what we are looking for in this technology platform.

Narrator: *A pause, and then everyone but Shalini starts talking at once. There's an apology, some awkward chuckling.*

DAVID: Sorry, Shalini, I guess you're seeing our excitement over fixing these issues.

CLAUDIA: Abra, why don't you start?

ABRA: We've been working on this for some time now. We started with an understanding of how processes are different from the functions and departments that we're used to, and there was some effort to get our heads in a different place. Once we embraced that, we went through a long effort to locate and name each of the processes. There were quite a few of them.

DAVID: Two hundred and thirty-nine—

ABRA: (*cutting in*) The actual number was a little different, but it doesn't matter.

CLAUDIA: I remember how we debated about what we can do to prioritize the processes.

ABRA: We were able to identify those high-priority processes—the high-value, often customer-facing processes that weren't working well—and that's what we've focused on.

SAMANTHA: We had help from some very capable staff to build and analyze these process models.

SHALINI: Not many companies go about this so systematically. I'm impressed.

ABRA: It has been quite a journey. Probably the most important thing we've done so far is our most recent step. We had teams working on those high-priority processes—they analyzed each process and came up with a redesigned process.

DAVID: So . . . now we find ourselves here.

SAMANTHA: Yes, and now that we're here, we need to find a way to figure out how to actually use all the work we've done so far.

Narrator: *Everyone turns to look at Shalini. Shalini slowly gets up and walks around the table, handing everyone a short, two-page, glossy brochure. As she returns to her place at the table, she remains standing. She has set the tone and will lead the discussion now.*

SHALINI: If you haven't heard about Sonita before, we are one of the leaders in the BPMS market, confirmed by our position in Gartner's magic quadrant for a while now—the handout I shared has some of these details. I'll share more about where we're going . . . and more importantly, how we can work with Royal Energy.

Narrator: *Shalini launches into a brief summary of the features of their software, and the group listens. Abra seems somewhat mollified. They do have plenty to offer. As Shalini winds down the general presentation, she turns to specifics for Royal Energy.*

SHALINI: To work with the foundation you have already built—which isn't bad, by the way—the first thing that comes to mind is that we now need to find some way of adding specific details for each task in these processes. I've heard about the interconnect process one of you mentioned in our email dialogue. To roll out the redesigned process, we have to now add those details, who is going to do each task.

DAVID: Yes! So, we have a task—I think for scheduling the technician—that needs someone to access the exact location of the property using GPS and other software. But at some point, I feel that we need to automate this task.

REGINA: And there are many others. I agree, we need each task specifically assigned to whoever is going to be responsible for that. We're actually doing this now, but that's just based on what we used to do. If we change the process, then we have a better way of ensuring that the right task is delivered to the right person at the right time.

SHALINI: And for each application.

DAVID: I thought it was for each customer.

SHALINI: *(leans over and wordlessly points to his printout)*

DAVID: Oh yes, right there . . . on the page . . .

CLAUDIA: We also need to find some way of figuring out who is going to do what task—adding specifics like an alert for that person on what task they need to do. Can they get an email or a text message, something like that?

SHALINI: With our software, you can do that. And you can generate a message to notify each person, such as "The next task for Application 12 is ready to be performed." In fact, you can set up our software such that the message will be delivered to whoever has a certain *role*—so that if the person in that role changes, it can still go to the right person. Whoever has that role, I mean!

ABRA: Like OneMedical.

CLAUDIA: Like . . . like what?

ABRA: OneMedical is sort of like a membership program for patients. You pay an annual fee and then you have access to all these clinics. Anyway, your doctor can give you prompts through the app to finish items on your to-do list. And all of these items are linked to the phone numbers and websites you need to complete each one. You can even send your doctor messages in the app and have virtual appointments.

SHALINI: Basically. Someone may have different responsibilities in different processes. They can get the right prompts so they can do what they need to do for each process.

REGINA: (*whispering*) Samantha, do you have an aspirin? Or four?

SAMANTHA: You know I do. Here. (*pills shaking*) You alright?

REGINA: I got a mean headache. I'll be alright.

CLAUDIA: Abra, that's a great example. (*lowered voice*) Also, can you text me that website?

SAMANTHA: That brings up another necessity: incomplete tasks. We have some things that sometimes get stuck in Operations. Once it moves there, we lose visibility, and it is so frustrating. I realize—now that we've done all these analyses—that they're working at almost full capacity. But this is still important. If a task is incomplete for some time, is there some way of checking on that? That way we can move that task to someone else if it's not completed . . . or at least pacify an irate customer.

SHALINI: Yes. You can set parameters like how long a task has remained incomplete to trigger an alert to the person responsible. You can also have a trigger to alert the supervisor or, in an extreme case, move the task to someone else.

DAVID: Let's also see it from an employee's point of view. I mean, what if they get lots of emails from all these different prompts from the process? That'll be way too overwhelming. Is there a way of prioritizing some emails over others?

REGINA: All of these are excellent points. Now, Shalini . . . does your software do all that?

SHALINI: Our software will do all of that and more. So far, we've all been talking about getting each process deployed and keeping it running . . . but once you roll out the processes and they start "running," you can do much more.

REGINA: Like what?

SHALINI: You can monitor the health of the process. You can see a dashboard that shows things like the number of cases that are completed on time, delayed, canceled, aborted; you can see these statistics in the aggregate; *and* you can see this for the current period as well as historical comparisons. It really gives you a window into how each process is performing.

REGINA: Great, great! But what about a particular customer? They're not worried about the completion percentage for the process; they want to know how their own request is being handled, right?

SHALINI: Yes, you can also examine particular cases. For example, if you want to look at Ms. Jones's interconnect application request, you can query for that and zoom in to look at the exact status of her request and let her know where things are. Based on the data, you'll be able to make very clear statements. For example, you can say things like, "On average, the interconnect process takes about fifteen days. Your request process is about 80 percent done and it's been only eight days. That means your request is well ahead of the average time it takes."

REGINA: If we can get that kind of visibility, our Customer Service folks would be happy. And our Operations Department will be able to make a clear case for their capacity constraints.

DAVID: So we can tell our customers exactly what's going on with their specific requests? Well dang, why haven't we done this the whole time? What you're telling us is that this is possible—and you have software that can be configured to do all of this.

SHALINI: You can even pull out some cases and push them ahead of others. So not only looking at a specific case. For example, imagine the possibility that someone in your Sales group can reach out to the right person in the Operations group right away—either via our software or informally—to prioritize some customers based on the importance you attach to that account.

ABRA: What's the price on this?

SHALINI: That depends on the number of seats, of course.

SAMANTHA: Do you have a hosted solution?

SHALINI: Generally, you want the solution to be deployed within your company so you can connect to your other systems. The users can authenticate against an LDAP server, for example.

DAVID: Can you provide any pointers to how you're positioned against other vendors?

SHALINI: As I shared earlier, we've been right up there in Gartner's magic quadrant for BPMS for a while now—in terms of both vision and execution. If you take a look at your handout, it has the graphs. I can share more about where we're going, if you need—like adding more capabilities for dynamic case management and robotic process automation.

REGINA: It looks like we already have several things that your software has . . . what's it called . . . BPMS?

CLAUDIA: I think . . . maybe what we're asking is if it's all one big software or if it has different capabilities that we can choose. Will it take the process models we've created so far?

SHALINI: As long as the models you are creating follow the industry standards, we can work with what you have. And based on what you've shared with me so far, your BPMN models look decent. We can deal with much larger models and are able to work with them quite efficiently, including monitoring the processes that run from these models. Your work so far may be able to flow right into the BPMS.

DAVID: BPMS?

SAMANTHA: Business process management system, David.

DAVID: Right, right. Thanks, Samantha. Sorry, Shalini, I'm the slow one.

CLAUDIA: I think this all sounds great, Shalini . . . it's good that the software is so flexible.

SHALINI: And it will give you a chance to be flexible too. For example, if one of your employees is promoted or leaves, we can swap out their email and attach a new one to the role responsible for the task. Then the inbox for that role will transfer over to the new employee. And as I shared earlier, we have a dashboard so you can see which processes are working well and which ones are not.

ABRA: Just a reminder that it's about time to wrap up.

SAMANTHA: There are a few things I still want to talk about—things I don't think we have good answers for yet.

SHALINI: Be my guest.

SAMANTHA: First, how much retraining do we have to do? Will our employees have to learn new ways to carry out their tasks?

SHALINI: Of course, there will be a certain level of that—but the idea is that the tasks will arrive with easy-to-follow prompts right into your email inbox. If you choose to click the link and work on your task, you can do that. Or you can also log into a mini-dashboard that each of your employees can have on their own computer.

ABRA: And that's talking about *your* software, BPMS?

SHALINI: (*chuckles*) I mean, yes, that's what we've been talking about.

DAVID: And that stands for "business process management system" . . . did I get that right?

SHALINI: That's right.

ABRA: Is there any other kind of software we can consider other than BPMS?

SHALINI: Well, nothing that has all the capabilities integrated like a BPMS.

REGINA: I hate to end this, people, but we've got to wrap it up. I have another meeting, two espresso shots, and more aspirin calling my name.

SAMANTHA: Okay, okay, but I'd really love to be able to try this for a few of the processes.

REGINA: We'll meet on this again, Samantha; I gotta run.

SAMANTHA: I do too. It was great meeting you, Shalini.

SHALINI: Pleasure's all mine.

(*everyone leaves but Claudia and Abra*)

CLAUDIA: Hey there . . . you alright?

ABRA: What do you mean?

CLAUDIA: Well, you were a little prickly with Shalini earlier. What's going on?

ABRA: I wasn't prickly . . . I wasn't the *one* who was prickly.

CLAUDIA: Ah . . . Shalini is very . . .

ABRA: Arrogant.

CLAUDIA: I was going to say confident . . . but that works too.

(*pause*)

CLAUDIA: They've got a good product—and she explains it well. What's the deal?

ABRA: I actually did like what I was hearing during the meeting, but . . .
               I'm frustrated.

CLAUDIA: No kidding.

ABRA: I wanted to finish my company software project before we had to
               include outside vendors.

CLAUDIA: Abra, I didn't know it was that important to you. Software engi-
               neering isn't even your main job, it's a hobby—

ABRA: A person can be a genius at more than one thing, Claudia.

CLAUDIA: Fair enough. But I'm sorry that you felt we didn't take your work
               into consideration. You're a valued member of this team, Abra.

ABRA: I appreciate you saying that. I'm not going to stop working on the
               software, but I will check my attitude at the door during our next
               meeting. (*goes to leave, but turns back*) Oh, and Claudia . . . thanks
               for checking in.

CLAUDIA: Like I said, we value you here, Abra.

To be continued . . .

Reflection Questions

1. At one point in the scenario, we hear Abra say, "Ah, so we *are* trying to 'throw technology at it.'" There appears to be a reaction to that from Claudia: "Uh, this is a little different." What do you think? Is that what Royal Energy is doing—throwing technology at the interconnect process? Explain.

2. You may be familiar with the company Adobe, which makes the Adobe Acrobat software. Other capabilities the company has created include e-signing and the ability to route documents in just the right order. See how they describe this functionality here: https://acrobat.adobe.com/us/en/sign/features/document-workflows.html. What is the software platform that Shalini is describing? Is it the same as this document-routing capability that Adobe promises? If not, what are the additional capabilities that Shalini promises that bring a set of benefits to Royal Energy?

3. In addition to the customer service issues that have been brought up several times in our scenarios, Royal Energy must also acknowledge that they are in the utility industry and, therefore, subject to regulations. This was not a point that came up during the conversation (Regina and David were in a hurry to wrap things up). Can the software that Shalini describes help Royal Energy deal with compliance issues any better? Explain.

Self-Assessment Questions

1. Can you describe the need for a BPMS for deploying business processes?
2. Can you explain the need for the different components in a BPMS?
3. Can you describe how a BPMS addresses commonly occurring problems?

Readings

1. Shaw, D. R., C. P. Holland, P. Kawalek, B. Snowdon, and B. Warboys. 2007. "Elements of a Business Process Management System: Theory and Practice." *Business Process Management Journal* 13 (1): 91–107.

2. Chang, J. F. 2016. *Business Process Management Systems: Strategy and Implementation.* New York: Auerbach.

Key Terms

BPMS (Business Process Management System)
A BPMS is a software solution that includes tools to design, model, execute, and monitor business processes. It has a number of components such as an execution engine, worklist handler, model repository, and many others.

Execution Engine

An execution engine in a *BPMS* is responsible for interpreting the *process model* for each *process instance*. It assigns each *task* to the appropriate *actor* and ensures that tasks are executed in the correct sequence based on the *sequence flow* and *gateways* specified in the process model based on the data in each *process instance*.

Execution Logs

The execution logs in a *BPMS* are records of activities that occur during the execution of a business process. They include a wide range of information, such as timestamps, task execution details, actor details, task state, and any problems or exceptions. They provide an audit trail that can be monitored, queried, analyzed, and examined to diagnose problems.

Process Model Repository

A process model repository in a *BPMS* is the central storage location for all process models in an organization. It allows different stakeholders and software tools to access and manage process models for purposes such as execution, version control, and more.

Task List

A task list is generated by a *BPMS* that organizes and displays the *tasks* assigned to a specific *actor*. It provides a real-time view of the tasks that an actor needs to complete as part of the various processes in which they participate. It can be used to notify actors, prioritize tasks, inquire about status, delegate responsibility, track progress, and much more.

Worklist Handler

The worklist handler in a *BPMS* manages the assignment of *tasks* to *actors* based on criteria such as workload balancing, task priority, and other business rules. It ensures that tasks are assigned to the right actors at the right time.

Scene 8
Buying Processes

Can't we just buy processes from vendors who design?
Use their software and retrain our employees to fall in line?

Learning Objectives

- Explain the options for acquiring processes
- Describe how enterprise systems can provide prebuilt processes
- Illustrate how an enterprise system can support business processes

Characters

Regina Wood—VP, Eastern MA
Anik Malik—VP, New Hampshire
Samantha Bellman—VP, Western MA
Sophie Raymond—Operations, Eastern MA (standing in
   for David Ashley, VP, Rhode Island)

Narrator: *The top dogs are meeting at the restaurant on the top floor of the Regency Hotel in Cambridge once again to discuss some options. Over the last several weeks, there has been increasing recognition of the efforts needed to follow a "homegrown" effort for process modernization. The group is considering the possibility of "outside" help but is not sure about how that can happen. At this meeting, everyone had expected David to attend. However, they're in for a surprise . . .*

(Sophie enters)

SOPHIE: Good morning, Samantha. Anik.

SAMANTHA: Good . . . morning?

ANIK: Sophie, what a pleasant surprise. What brings you here?

SOPHIE: . . . So David didn't tell you.

SAMANTHA: Tell us what?

SOPHIE: I am standing in for David today. He had something come up, and—

ANIK: That does it, he's always doing this!

SAMANTHA: Sophie, do you know if he's going to show up at all? We have some important decisions to make.

SOPHIE: I'm sorry, I—

ANIK: No, no, don't be sorry, Sophie. It's not your fault. David misses meetings every now and then, but he's never sent someone in his place. (*phone dings*) Hold on, I'm gonna respond to this.

SAMANTHA: (*irritated*) You may as well have a seat. We ordered for David, but please feel free to order something else.

SOPHIE: Oh, no . . . I'll have whatever—

REGINA: Well, well, well, if it isn't the old gang. Oh, hi, Sophie! What, uh . . . ?

SAMANTHA: David didn't show. He sent Sophie.

REGINA: (*sigh*)

SAMANTHA: I know.

SOPHIE: If it's not good for me to be here, I really can just—

REGINA: No, no. I'm glad you're here. You are always well prepared. We appreciate that. We look forward to your insights. Please, stay.

SOPHIE: Alright. Thank you.

SAMANTHA: I vote that we just go ahead and start before the food gets here. With some luck, maybe we'll finish the meeting and have time to enjoy the meal afterward.

REGINA: Sounds good to me. And we have a lot to discuss. Although we say this every time we meet, let me repeat: we've been working on this project for several months now and have gotten so many people involved, especially our chosen teams.

ANIK: I must say I've been very impressed with the work that Abra, Claudia, Kevin, and Jamie have done with identifying and naming their set of processes. Remember how Claudia found someone to go through that huge catalog? Her team really put the work in on that one.

REGINA: I'm glad we got that information.

SAMANTHA: And remember how that team had to learn to prioritize the processes they found? How many was it again?

ALL:	Two hundred and thirty-six.
SOPHIE:	. . . Roughly.
ANIK:	At first we couldn't get that number right; now it shows up in my dreams.
SAMANTHA:	You should talk to someone about that. (*Anik's phone dings*) You're a busy bee today.
ANIK:	It's nothing, it's fine.
REGINA:	I think a turning point was learning how to model and analyze the processes. Sophie, I never got a chance to tell you what wonderful work you and Remy did. How is he doing these days, by the way?
SOPHIE:	Oh, thank you. Um . . . Remy and I haven't spoken much since our part of the project wrapped up. I think he planned to go back to Spain to spend some time with his family.
ANIK:	(*spills coffee*) Oh, come on . . .
	(*simultaneously*)
SAMANTHA:	Oh my gosh! I'll call the waiter.
REGINA:	Did you burn yourself?
SOPHIE:	Here's a napkin.
ANIK:	No, no, it's alright. I'm fine, but this is cashmere, so I have to rinse it off or Joe will—(*phone buzzes*) Jeez . . . I'll be right back. (*rushes out in a huff*)
REGINA:	Actually, I'm glad Sophie is here, because we're going to need her help and Remy's for the ongoing work of deploying the solutions we've come up with. Recounting all of this is good . . . it's important because it's taken a lot of teamwork to get here. And we've still got work to do.
	(*Anik returns*)
ANIK:	Sorry about that. The waiter helped me blot most of it off. Where were we?
REGINA:	Just discussing the next steps. (*Anik's phone dings*) Everything okay?
ANIK:	(*shortly*) It's fine. Family stuff.
SAMANTHA:	What we've done so far has taken a ton of effort and resources, and I know we're not finished, but I'm wondering how on earth we're going to do this for every process we have.
REGINA:	Well, we have BPMS—can't we use that?

SAMANTHA: Of course, but that only helps us deploy the processes after we've modeled and analyzed and redesigned each. Sophie, how long did it take for you and Remy to redesign that one process you were dealing with?

SOPHIE: It took a while, Samantha's right. And to get a well-designed process, we first have to model the process as it currently is, analyze it, find ways to improve it, then come up with a model of what we improved. We need to collect historical data about how the process is performing, work with several stakeholders—it's not easy to do.

SAMANTHA: Sophie, you're doing a great job—but imagine doing that for all the processes we care about. I know we've had teams working on the high-priority processes, so we should have made better progress than—

REGINA: Samantha, we're all just as worried about the sheer scale of this as you are. I promise.

SAMANTHA: (*huffs*)

REGINA: So . . . doing all of this for every process takes a long time. What is the alternative?

ANIK: And are there any alternatives? Do we have any real options?

                (*pause*)

SOPHIE: I . . . I might have a solution. Well, David has one. He, well, "we" prepared it and did the research and typed it up. Um, here are some . . .

Narrator: *Sophie and David have been exploring another direction. They have a concrete idea that they want to share with the group. This is precisely why Sophie is here today. She passes a short write-up to everyone around the table.*

REGINA: What's this?

SOPHIE: We knew that the effort involved in analyzing the processes was significant, and it has been taking too long. David asked me to look up some options. He and I had been discussing it, and we took the liberty of contacting your chosen team leaders to ask them to come up with some solutions.

ANIK: So you're asking us to change gears now? We have had several teams already working with many processes—what do we do with all of that effort? Throw it away?

SOPHIE: (*gingerly*) There has been an investment of time and effort. But that's all sunk cost, isn't it?

Narrator: *The group is listening intently. Sophie has just brought up a trigger that they have themselves used when talking with stakeholders. Whatever they have invested in the old way of doing something, the old process, is already spent—it is a sunk cost; it should not keep them from moving toward a redesign. Well, they cannot run away from this now. She has the floor.*

SOPHIE: (*clears throat*) So, there are a few options. The first one utilizes the value chain—

REGINA: Value chain? I remember learning this back in the day. Anik, don't you remember—

ANIK: I do. That's a fundamental concept and, actually, a perfectly logical place to start.

SOPHIE: Yes, the idea comes from research done in the eighties, but it's still clearly relevant. Basically, Royal Energy can start by using a value chain and examine each cluster. Logistics, Operations, Marketing, Sales . . . all these are the primary activities. Once we examine them, we can identify the processes within each cluster.

REGINA: This would allow us to identify processes for, say, procurement, such as purchasing, stocking, vendor assessment—

SOPHIE: Yes, and so on! Or even processes for human resource management such as hiring, performance appraisal, firing, benefits, etcetera.

REGINA: Well, at least we recognize this one.

ANIK: True. I agree with all of this, but it doesn't solve the problem, does it? We did some of this before about identifying processes with a different approach. Now we still have to redesign the processes.

SAMANTHA: And I don't see how this idea is going to set us apart from the younger companies who are running ahead of us.

REGINA: Well, sometimes we have to look back so we can move forward. Any other options, Sophie?

SOPHIE: Yes, as we mentioned earlier, Claudia was a big proponent of at least considering a process catalog. These can be actual books and binders that define the terminology, the overall set of processes, and the KPI for a company operating in an industry. They are the—let me read this—"collected wisdom of consultants and experts in an industry."

SAMANTHA: Two questions: First, can you buy these catalogs, and if so, how do we move from pages in a book to actual deployment? And second, how is this moving forward? This seems just as difficult as the value chain or doing our own processes. We have to study and understand this entire book or catalog?

ANIK: Sam . . .

SAMANTHA: Alright, alright.

SOPHIE: Claudia, who did the research on the catalogs, is aware that this is an older way of working on things, but she thinks using a process catalog could be really useful. They come in two different versions: industry-agnostic and industry-specific. So there are process catalogs for airlines, banking, city government, education, health care, retail, utilities—

REGINA: That's us!

SAMANTHA: Okay, this sounds a bit more promising, though I'd be interested to start with the general version . . . that's the agnostic version, right?

SOPHIE: Right . . .

SAMANTHA: Yes, I'd be interested to start there so we can refine a generic process how we want.

SOPHIE: And that's totally possible. These process designs may be seen as old-fashioned, but they represent the collective wisdom of APQC over the last twenty-plus years. A company can select whichever process they want and apply it to their organization.

ANIK: Uh . . . what's APQC again?

SOPHIE: APQC is (*reading out*) American Productivity and Quality Center. Their website describes the organization as "the world's foremost authority in benchmarking, best practices, process and performance improvement"—and so on.

SAMANTHA: So they've been doing this for a while . . . what version are they on?

SOPHIE: Sorry?

SAMANTHA: For utility companies. They have been building these process catalogs, right? How many versions of process catalogs have they gone through? So we can get a sense of how mature their thinking is . . . I will assume they have a catalog for the utility industry, am I right?

SOPHIE: Oh, um (*rustles papers*), I believe they are on version seven or eight of their *Process Classification Framework for Utilities*. I may be wrong,

	though; I will need to check. All I am saying is that this is not fresh or untested.
REGINA:	That's great. That means it's gone through several lives and would prove useful for us. (*sound of cell phone keys tapping*) Anik, what do you think?
ANIK:	. . .
REGINA:	Earth to Anik . . .
ANIK:	(*forceful sigh*) I'm sorry, guys, I have to . . . make a . . . (*answers phone quietly while moving away from the table*) Babe, come on . . .
SOPHIE:	Um, should I . . . ?
SAMANTHA:	No, just keep going.
SOPHIE:	Okay, well, I was going to say that since Royal Energy is a mature company, they're the perfect candidate for this option. But there is a third option. It's more complex, but it's a modern approach. Samantha, I think you'll like this one.
SAMANTHA:	(*reading through the handout Sophie had shared earlier*) You're referencing VRM and SCOR?
SOPHIE:	Yes! Value reference model could be used for—
	(*Anik returns*)
ANIK:	Thanks for your patience, everyone.
REGINA:	What's going on?
ANIK:	Our dog got caught in a glue trap. He's at the vet.
SOPHIE:	Oh no!
ANIK:	He's okay, this happens a lot. He's old, but he's always been an idiot.
REGINA:	Poor little thing.
SAMANTHA:	Yes, poor baby. Anyway . . . Sophie was just reminding us about VRM. Anik, what do you think of the options of VRM and SCOR as reference models moving forward?
ANIK:	I think we should consider the SCOR model.
SOPHIE:	They are useful, but very generic. We may want something more focused on utility companies like ours.
SAMANTHA:	Ugh.
REGINA:	But what's the real difference between, say, APQC's suggestions and SCOR?

SOPHIE:	SCOR helps you balance and manage things like inventory levels, risks, and goals across the range of your suppliers so that you can better optimize the complete supply chain. As far as I can tell.
SAMANTHA:	As far as you can tell?
SOPHIE:	Well—
SAMANTHA:	I'm just being grumpy, ignore me. I just want SCOR to work for us. I want *something* to work for us.
REGINA:	Can't we combine VRM and SCOR somehow?
SOPHIE:	It's currently being researched, but it hasn't been done yet. There's no method or tool available to make these two options compatible.
SAMANTHA:	What about vendors? There have to be vendors who have implemented the processes we need. Then we can just buy them and roll them out, right?
SOPHIE:	(*relieved*) I'm so happy to hear you say that, Samantha! Yes, and that's exactly our option four: enterprise systems that already include implementations of what are considered well-designed processes.
SAMANTHA:	Oh, right . . . we've discussed this. There are vendors who have done all this work for us; they have redesigned and implemented the processes, right?
SOPHIE:	Yes, the only caveat is that they won't be specific to Royal Energy.
ANIK:	So back to the generic stuff?
SOPHIE:	No, no. They have a version that's tailored to the utility industry—
REGINA:	Yes, so industry-specific versions of the ERP systems, is that right?
SOPHIE:	Right, that's Jamie and Abra's research. There are vendors who have these software modules with prebuilt processes . . . (*a little more forcefully*) and they have a version that's tailored for companies in the utility industry.
SAMANTHA:	I'm waiting for the *but* . . .
SOPHIE:	(*smiling*) . . . But these are processes that may not be exactly like how we do stuff—and that means we will have to train our employees in each of the divisions.
ANIK:	Well, I guess the benefit is that we'd get prebuilt processes?
SOPHIE:	Correct. And what's important is that these processes will be well designed—they may not be superinnovative, but they would be efficient.

REGINA: But they won't be Royal Energy–specific and we'd have to retrain everyone.

SOPHIE: Also correct.

SAMANTHA: There are some companies that have done this successfully. SAP is one of them.

REGINA: That German company? Oh, I forgot that your dad—

SAMANTHA: Yes, my dad is obsessed with that characteristic German thoroughness.

SOPHIE: They can recycle anything over there, I've heard.

SAMANTHA: (*laughs*) Yes, almost everything.

SOPHIE: There are a few other companies who deliver ERP systems—there's Oracle, Microsoft, and some others as well.

SAMANTHA: These are big established vendors, so this seems like a viable option. Maybe it will work for us.

REGINA: (*claps hands and rubs them together*) Okay, so . . . I will guess that the way this works is that we buy that software and deploy it.

ANIK: But it will be hard to do that for us. We'd have to make a lot of changes.

SOPHIE: True—but let's remember, we will buy a version of their software with modules and processes that are built for the utility industry.

ANIK: What exactly does that mean?

SOPHIE: Well, there will be support for what Royal Energy cares about—like managing large infrastructure assets, initiating service for customers, managing billing and revenues from our customers, scheduling technicians, and so on—and not for things like inventory optimization and distribution logistics that a big-box retailer may need.

ANIK: Oh, interesting. So they build totally different versions of their software for each industry?

SOPHIE: Their software has these things they call modules. Some modules are common like financials and HR.

ANIK: But . . .

SOPHIE: But other modules are specific to each industry. They have a set of modules that are designed for companies in the utility industry.

ANIK: And that's what we would want.

SOPHIE: Yes! Of course, they have other modules for the retail industry, some for the higher education industry, some for professional services, and so on.

ANIK: Got it.

SAMANTHA: And the processes, they are built as part of those modules?

SOPHIE: Yes! Some of them are implemented as part of a single module; others may run across modules. Each module may be run by a department in Royal Energy, like Finance, Warehousing, HR, and so on.

SAMANTHA: (*suddenly brightening*) And they have prebuilt steps that different employees in different departments must take to carry out a process.

SOPHIE: Yes—that's why I said these will be efficient processes because they have designed and refined these and implemented these as part of their software.

ANIK: That means when we buy their software—

SOPHIE: The specific modules from their software offerings.

ANIK: Yes, the specific modules—we are actually buying those prebuilt steps that run across modules. But the downside is that those pre-built steps may not be exactly how we do things at Royal Energy.

SOPHIE: I couldn't have said it better. So we will short-circuit this already heavy lift we have of examining, analyzing, and redesigning and rolling out improved processes.

SAMANTHA: But the downside is that these will be new processes, and we will have to train a lot of people.

ANIK: So we won't be just buying this—what did we call it—enterprise software? ERP system? We will be buying into a way of doing business.

REGINA: Yes! That is a point worth repeating. So they'd better be good at this!

Narrator: *The discussion so far has provided everyone a glimpse into a new future for Royal Energy, one that includes a very strong influence from an external vendor who will essentially dictate how most processes will be run at Royal Energy. Regina returns to the handout from Sophie and starts reading it. Anik can't stop muttering. Samantha is excited but cautious. Sophie senses the shift in the mood. The group has realized they are on the verge of making some decisions that will change the trajectory of how Royal Energy will operate for the future, and in such a profound way that it will be impossible to go back! What started as a normal* *meeting has suddenly turned into one where everyone has realized the importance of the decision facing them.*

SOPHIE: *(breaks the silence)* Uh . . . I suggested to David that we should get the department heads and the vendor representative in the same room to discuss these things.

REGINA: An excellent idea.

SAMANTHA: Clearly we'll need to get the process experts in on this meeting as well.

SOPHIE: Well, yes, that would be—

ANIK: But Remy isn't available, right?

SOPHIE: I'll give him a call and see.

SAMANTHA: Let's get this together quickly, folks. Sophie, you have given us much to think about. So what now? Adjourn?

REGINA: I think so, yes. And just in time for the food to arrive!

Narrator: *The group tucks into their food. Business is set aside in favor of friendly banter. Even as they relax, the weight of what they have learned lingers. Even Anik, distracted by his personal issues, cannot shake this sense. At the end of the meal, the group gathers their things to leave. Samantha checks in with Sophie.*

SAMANTHA: *(still pondering what they have discussed today)* Sophie, thank you for your prep today. We would not have been able to have this conversation so quickly if you and David hadn't prepared. But I need you to be honest about something.

SOPHIE: Okay . . .

SAMANTHA: You did most of this prep, didn't you?

SOPHIE: . . . Yes. But David's a good coach.

SAMANTHA: Well, you made it happen. Good work.

SOPHIE: Thank you.

To be continued . . .

Reflection Questions

1. As the group starts to consider "outside" options, Sophie points out (paraphrased here), "Claudia was a big proponent of at least considering a process catalog. These can be actual books and binders that [reflect] the . . . 'collected wisdom of consultants and experts in an industry.'" In response, Samantha asks, "Can you buy these catalogs, and if so, how do we move from pages in a book to actual deployment?" Do you think Royal Energy can benefit from such process catalogs? Why? Why not?

2. As the conversation proceeds, the group considers another option (with some caution). As Samantha notes, "There have to be vendors who have implemented the processes we need. Then we can just buy them and roll them out, right?" This time, Sophie responds, "Yes, and that's exactly our option four! Enterprise systems that already include implementations of what are considered well-designed processes." Do you think this is a feasible option for Royal Energy? If you were a manager of one of the departments (like Finance or HR) at one of the divisions of Royal Energy, how would you react? Explain.

3. The group appears to go through two long cycles of discussion about this idea of enterprise systems. Sophie explains, "We will buy a version of their software with modules and processes that are built for the utility industry," and then Samantha articulates, "And they have prebuilt steps that different employees in different departments must take to carry out a process." What is the sudden realization that dawns on the group based on this discussion? If you were the VP for one of the divisions at Royal Energy, what would you do with this newfound understanding?

Self-Assessment Questions

1. Can you explain the options for acquiring processes?
2. Can you describe how enterprise systems can provide prebuilt processes?
3. Can you illustrate how an enterprise system can support business processes?

Readings

1. Gupta, M., and A. Kohli. 2006. "Enterprise Resource Planning Systems and Its Implications for Operations Function." *Technovation* 26 (5–6): 687–96.

2. Magal, S. R., and J. Word. 2011. *Integrated Business Processes with ERP Systems.* Hoboken, NJ: Wiley.

Key Terms

Enterprise System

An enterprise system (sometimes called an enterprise resource planning [ERP] system) is a type of software that organizations use to manage day-to-day business activities. It provides predesigned processes (e.g., sales order process) built on top of modules (e.g., accounting, inventory) that store the data. This integration of processes and data provides a turnkey software solution for an organization. Examples of widely used enterprise systems include SAP, Oracle, and Microsoft.

Industry Versions (of an Enterprise System)

An industry version of an *enterprise system* is one that is designed for the industry vertical. For example, an enterprise system for higher education may include processes such as class scheduling and faculty evaluation, with modules such as program management and alumni relations. On the other hand, an enterprise system for the retail vertical may include processes such as procurement and sales forecasting with modules such as inventory management and store management.

Master Data

Master data are shared and used by multiple processes and applications in an organization. Examples include customer information, vendor information, material master data, and asset master data.

Modules/Components (in an Enterprise System)

Enterprise systems typically consist of *modules*, each built on a business area (such as Finance, HR, Marketing). Organizations often acquire the *industry version* of the *enterprise system* appropriate for their organization and, after that, a subset of the available modules based on their specific requirements.

Organizational Data

Organizational data describe the organization and its operational structure. They are typically defined during the initial setup and rarely change. Examples include company codes, business areas, company locations, and work centers.

Transactional Data

Transactional data are generated from business operations, such as the consequence of every transaction. They are created and stored to document the transaction. Examples include sales orders, invoices, purchase orders, payments, and shipments.

Scene 9
Customizing Processes

You may need to configure, customize, polish edges
Be prepared to pay the consultants some high wages

Learning Objectives

- Explore options for modifying the prebuilt processes
- Evaluate the options with criteria appropriate for enterprise systems
- Describe the feasibility of each option in terms of internal and external resources

Characters

Kevin Sanders—Scheduling/Dispatch, New Hampshire
Jamie Cochrine—Operations, Rhode Island
Claudia Narez—Environmental, Eastern MA
Abra McGregor—Finance, Western MA
Remy Garcia—IT, New Hampshire
Sophie Raymond—Operations, Eastern MA
Matthias Schuster—TBQ, an Enterprise Systems Vendor

Narrator: *In the last meeting, the group considered the possibility that one can buy prebuilt processes instead of examining, analyzing, and redesigning each process with a heavy investment of time and effort internally. There was a realization in that meeting that these prebuilt processes—even though tailored for the industry segment—may still be too different from how Royal Energy operates. Some in the group have expressed concerns that these prebuilt processes will simply not fit the way things are done at the company. The meeting today is meant to explore what's feasible in terms of modifying these processes. It's been difficult to get everyone together. In the meeting today, we have the department heads from each division as well as Sophie and Remy, the latter having postponed his sojourn to Spain for the sake of this meeting. And there is a visitor . . . Matthias, an enterprise systems vendor representing TBQ. Matthias's personality can be boiled down to two adjectives: cheerful and weird.*

CLAUDIA: Kevin, have you met, um, how do I pronounce your name again?

MATTHIAS: Oh, it's not a problem at all, really, um, Math-ee-us. That works.

CLAUDIA: That works . . . but is that how you pronounce it, or is there another way you—

MATTHIAS: Oh, nobody ever gets it right, so it's really just whatever you want, really, it's just . . . yeah . . .

KEVIN: Why aren't there snacks here? I'm trying to figure out who dropped the ball.

JAMIE: Kevin, we all thought you were going to bring something because you're always talking about it.

KEVIN: You kids don't know the power of snacks in a meeting. That's just us, right, Claudia?

CLAUDIA: As usual, Kevin, I have no idea what you're talking about.

MATTHIAS: Oh, I have some licorice, it's really yummy, it's from . . . but y'all wouldn't want . . . I mean . . . never mind.

CLAUDIA: I say we get started. Everyone, this is . . . Math-ee-us. He's from TBQ.

(chorus of hellos)

CLAUDIA: I'm Claudia. This is Kevin, Abra, and Jamie. We've also got Remy and Sophie, who are our process experts.

(more hellos)

CLAUDIA: Just so we all start on the same page, I'll remind the group that our division heads asked us to set up this meeting to better understand what needs to be done—*if* we buy those processes—and *after* we buy those processes. For example, what our options are for modifying those processes to fit what we need at Royal Energy. Next steps, folks! We're moving forward!

(chorus of small cheers, mostly Matthias, who quickly feels awkward at clapping too loudly)

JAMIE: So, Matthias, we're planning on buying the TBQ enterprise system—the version that you have for the utility industry. Or I should say, we're *thinking* of buying the TBQ enterprise system.

MATTHIAS: Great! How can I help?

JAMIE: Well, so as I see it, when we buy your software, we're actually buying all those prebuilt processes.

KEVIN: Yeah, so . . . in order to get your software into Royal Energy, and have it working efficiently, what are the things we need to understand?

MATTHIAS: So, what you do is that you end up buying different modules—these are software modules like Finance, Manufacturing, Dispatch, and others—and each has processes that are prebuilt. Some processes are mostly run inside a single module; most run across modules even if they are seemingly controlled from a module.

Narrator: *Sophie and Claudia exchange knowing glances. This is what they discussed in the last meeting. It has been a few weeks since then, but the power of the ideas has not dissipated. Remy nods at Sophie, who has a satisfied smile on her face. She is pleased with herself; this vendor rep is using words similar to how she described enterprise software and the ideas of prebuilt processes.*

KEVIN: Yeah, but . . . and I might be getting ahead of myself here, but I'm assuming there's gotta be a downside to this. We sort of know what we're getting into, but not really.

MATTHIAS: That's where I come in, yay! Be at ease! Hehe, uh, there are a couple things I want to emphasize. We have a piece of software that's being used right now by several Fortune 500 companies.

ABRA: How many? A few?

MATTHIAS: Uh . . . many? It's quite a few. I can get a complete list if—

CLAUDIA: No, we don't need that. You can just keep going.

MATTHIAS: Oh, okay . . . heh, um, well, the software is quite reliable because it's gone through many versions. The software has some core modules, and that means any company that wants to buy their software first has to buy the core modules. And then there are other modules that surround the core that a company can pick and choose.

SOPHIE: Matthias, did you need to use the projector?

MATTHIAS: . . . I'm sorry?

SOPHIE: The projector. We had it brought in for you to—

MATTHIAS: Oh, jeez! Sorry I'm such a diddlywinks. I completely . . . anyway, here's a visual description of . . . basically what I just said. Lemme just hook up my laptop and . . . ta-daaaa!

               (chorus of small agreements and sounds of understanding)

CLAUDIA: Okay, this is making more sense, but I have a couple questions. First—and I think I know the answer to this but just in case—you have different versions that are targeted to different industries, correct?

MATTHIAS: Yessss . . . yes, we have a version for the utility industry, and that would definitely be the right fit for Royal Energy. And even within that version for the utility industry, you can pick different modules.

CLAUDIA: Can you give us some examples of what you mean by *modules*?

MATTHIAS: Um, well, yeah! Let's see, we have the core modules of course, like finance and HR, which we believe all companies will need. But Royal Energy will not need a module for, say, grants management that a university may need or a module for manufacturing or a module for inventory tracking in stores that a retailer may need.

CLAUDIA: Tell us what you think we might need.

MATTHIAS: (*continuing*) Oh boy, okay, so there's revenue management that you need, infrastructure and asset management, and you need dispatching and scheduling. I can point you to the complete list of modules for the utility industry. Aaaaaand . . . uh, do you have any other questions?

CLAUDIA: All we have heard so far is about modules. Where are the processes? Where do they fit in?

MATTHIAS: So glad you asked! (*clicks projector*) Processes are built on top of those modules. I'll give you an example. Y'all have a procurement process, right? So . . . say you have someone from warehousing who detects a lower inventory level on some items like the meters that you install. They would build a purchase requisition and send it to the purchasing department. The purchasing department will need to look up vendor contracts and issue a purchase order to the vendor. The vendor will eventually ship the items to the warehouse, the warehouse will prepare the goods-received document—oh, y'all have that, right? Goods-received documentation?

            (*small chorus of "Yes"*)

SOPHIE: Our people in different departments will use the different modules you have. When Eddie from Warehousing enters the purchase requisition, that will trigger something in your software. Your software already has all those built-in connections, so that will trigger the next step in our Purchasing Department to generate a purchase order. That's the start of the procurement process, isn't it?

MATTHIAS: (*beaming, almost relieved*) Yes! You got it!

CLAUDIA: Want to tell us a little more?

MATTHIAS: Oh, of course. So we talked about the goods-received document—now, *that* will be sent to the Finance Department. By then, they probably also have the invoice from the vendor. Aaaaand, next part . . . (*clicks*)

Now the Finance Department will compare the purchase order that you issued, the goods-received document, *and* the invoice the vendor has sent. If all three of them match, it's all good! Hehe . . . um, and then they'll release payment. You know what I'm talking about! Lend me some sugar! I am your neighbor!

Narrator: *Matthias's energy and excitement for his company's technology usually find a more receptive audience. However, our group isn't accustomed to his strange theatrics and is mostly silent. Sophie awkwardly offers a thumbs-up.*

MATTHIAS: Yeah, a thumbs-up is fine . . . okay, good, so *anyway* . . . all of this is happening with different modules. Our friend in Warehousing will use the inventory module from TBQ. Our friend in Purchasing, or maybe the purchase rep, will use the purchasing module from TBQ. And our friend in Finance will use the finance module in TBQ. The way we've designed the modules, all of them are wired together. So they all reach into the same centralized database. That means this procurement process can be run by different departments in different parts of the company using different modules—and TBQ will be responsible for keeping the data consistent across different steps of the process. The software will ping different people across departments to do different tasks! It's a really cool . . . thing. It's really cool. And the data are stored centrally and accessed from the different modules. And *that* is how all the processes are prebuilt. The end.

(someone claps fast and gives a "Woot!" [probably Kevin] and someone claps slowly [probably Abra])

JAMIE: So I feel like I can speak for everybody when I say that this procurement process you describe makes perfect sense to us. The tricky part is that we have so many variations of this. For some items, we trigger an order when the quantity falls below a certain point. For others, we have blanket contracts with vendors, and for some others, we put out a request for a bid every time.

ABRA: Yes, and for some items and with some vendors we have different tax structures to which we need to pay attention.

JAMIE: Right—so, how will we handle all of these with your software?

MATTHIAS: Ah, my friend, 'tis all good! *(clicks)* Each prebuilt process comes with options. Think of these as switches that you can throw on or off. For example, we have a switch that will activate a part of the process that puts out requests for bids. You turn that on when you have to do procurement for things where you request bids and leave it off when you don't need to. And then *(clicks)* there's another switch that will allow you to rely on a blanket vendor contract. You will find that the documentation and interface at TBQ are the best in the industry!

KEVIN: Wait, wait . . . so you're telling me that we could've just talked to you when we started this whole project and we could've just bought your software to begin with? And . . . and . . . we can do all these changes this way? So we can get fully operational processes that are gonna work specifically for us? Why didn't we start with this Matthias guy?

(everyone is listening to Kevin think his way through this; they are used to it by now)

KEVIN: So, Matthias, you gotta tell me—we can do all this, right? And there's no downside?

MATTHIAS: Ehhh, you can do a *lot* this way. We call this configuration. Now in some rare cases, you may have some things that we do not have switches for. Then, if you like, you can write custom code.

ABRA: Like with a programming language?

MATTHIAS: Yes, we have a specialized programming language that is geared to our software, just, you know, so it's easier to work with.

ABRA: Oh . . .

JAMIE: "Oh" is right. You have your own programming language? We'll need to find talent who can do that . . . I don't think our IT group will have that expertise.

ABRA: I don't think we have anyone in the company who can do it.

MATTHIAS: Well, my friends, you are in luck! TBQ has managed to grow a large ecosystem of developers and consultants out there. We run classes for them so you can locate and work with anyone you want. We certify them when they pass, so you have some confidence!

ABRA: How large?

MATTHIAS: *(laughs nervously)* Ha-ha, uh . . . what?

ABRA: How large is the ecosystem?

MATTHIAS: Oh! Uh . . . fairly? Um, of course, you can contract with TBQ as well. In that case, we can do this for you—or you can directly hire any of the external consultants.

KEVIN: Wow, this sounds like our pot of gold for sure!

(murmurs of agreement)

Narrator: *In spite of his goofy demeanor, Matthias's excited approach and command of the facts has been effective. It appears that the group is pretty much ready to give in and sign on the dotted line to bring in TBQ to Royal Energy. But Sophie's work has made her aware of problems with enterprise systems like the one from TBQ. She decides to interject.*

SOPHIE: Not to burst anyone's bubble, but I have to ask . . . if we do this configuration by throwing different switches and coding—

MATTHIAS: Customization. So when you do coding, it's customization. Sorry to interrupt, though!

ABRA: And we do that coding—customization—with your programming language, it's not like Java or something.

KEVIN: There you go, that's the downside. Because *we* don't have that talent, I'm sure—that means we have to either contract with TBQ or use one of their consultants.

SOPHIE: Right . . . configuration or customization . . . then will that solve the puzzle for Royal Energy? As I see it, we'll make these changes, modify the processes either by throwing switches to activate or deactivate options—

MATTHIAS: Right, configuration . . .

SOPHIE: . . . or hiring consultants to write code and modify the processes—

MATTHIAS: Customization.

SOPHIE: Right. And then we'll need to work on actually rolling these out into the different divisions we have. So the question I have is . . . what happens after that?

Narrator: *The group is listening intently now. They've seen Sophie grow quickly as one of the sharpest minds on the project, and today she's really standing out. The group doesn't fully understand what point she's trying to make, but they trust her direction.*

JAMIE: That's a good summary, Sophie. We will have to train our employees with these new processes, right?

          (everyone is following along now, nodding)

SOPHIE: Right. So, after we do all this work, which may take months of effort—what's going to happen when TBQ releases the next version of the software?

          (the group is stunned; they have not considered this possibility)

CLAUDIA: Uh . . . that's a good question! Matthias, what do we do then?

MATTHIAS: Uh, well . . .

KEVIN: Oh no, Matty . . . don't do this to me! I knew it—it's all too good to be true.

MATTHIAS: . . . *Most* of your changes should be good and will be carried forward from the last version, but you know . . . we can't promise that *all* your changes will move over seamlessly from one version of the software to the next—

REMY: So that means we'll have to do it all over again just because you release a new version?

Narrator: *So far, the team has been quite upbeat. Now with this revelation, the mood shifts. This is also the first time Remy has said anything since the meeting started.*

MATTHIAS: Well, uh . . . Remy, was it? That's sort of true, but companies have been known to do it. And after a while, you just get used to it.

                (*sounds of frustration*)

CLAUDIA: Well . . . this is good to know.

KEVIN: My heart is broken. Nothing is ever easy around here.

JAMIE: Wait, so you mean you would want us to retrain our employees and then bring in a new version of TBQ that takes away all that our IT group would have done to modify the processes?

CLAUDIA: You know what? Let's call it a day—this requires some more thinking. Abra, will you send us those notes?

ABRA: Yup, I always do.

CLAUDIA: Great, thanks. And thank you, Matthias, for showing us what TBQ can do for us.

MATTHIAS: Oh, it was my pleasure, really!

To be continued . . .

Reflection Questions

1. As part of this conversation, Matthias confirms what the group had heard from Sophie in the previous meeting: enterprise systems come with prebuilt processes. He describes how the procurement process can unfold with TBQ, saying, "The way we've designed the modules, all of them are wired together." Here, he seems to be saying something specific that distinguishes enterprise systems like TBQ from the other option the group has considered before, a BPMS like Sonita. What is this key difference and why might this be relevant for Royal Energy?

2. A little later in the conversation, Jamie says, "This procurement process you describe makes perfect sense to us. The tricky part is that we have so many variations of this." Matthias responds, "Each prebuilt process comes with options. Think of these as switches that you can throw on or off." This appears to be one specific approach to modifying processes. How would you characterize this? Should Royal Energy use this approach? What are the limits to this approach?

3. The conversation progresses, and Matthias, in his eagerness to showcase all that TBQ can do, describes another approach to modifying processes. He put it thus: "We have a specialized programming language that is geared to our software." This appears to be another specific approach to modifying processes. How would you characterize this approach? Should Royal Energy use this approach? What are the benefits/problems associated with this approach?

4. Finally, it appears that Sophie is able to add a healthy dose of reality to everyone's enthusiasm about TBQ by asking, "After we do all this work [modify the processes and roll these out], which may take months of effort—what's going to happen when TBQ releases the next version of the software?" What should Royal Energy be prepared for? Is this a proverbial "showstopper" that will prevent Royal Energy from moving forward with an enterprise system like TBQ with its prebuilt processes?

Self-Assessment Questions

1. Can you explore options to modify the prebuilt processes for your organization?

2. Can you evaluate these options with appropriate criteria?

3. Can you describe the feasibility of these options in terms of resource needs?

Readings

1. Brehm, L., A. Heinzl, and M. L. Markus. 2001. "Tailoring ERP Systems: A Spectrum of Choices and Their Implications." In *Proceedings of the 34th Annual Hawaii International Conference on System Sciences*, vol. 8. IEEE Computer Society Digital Library. https://www.computer.org/csdl/proceedings/hicss/2001/12OmNAkEU3s.

2. Hustad, E., M. Haddara, and B. Kalvenes. 2016. "ERP and Organizational Misfits: An ERP Customization Journey." *Procedia Computer Science* 100:429–39.

Key Terms

Configuration

The configuration of *enterprise systems* refers to setting system parameters and options to bring the functionality of the enterprise system closer to the organization's requirements and practices. It involves setting system preferences and switches for different functionalities.

Customization

Customization involves modifying the software code of the *enterprise system* to add new features or change its functionality. It is more time-consuming, costly, and risky compared to *configuration*. It can also make upgrades more challenging. Therefore, most organizations try to use *configuration* as much as possible and resort to customization only when necessary.

Scene 10
Rolling Out Processes

Buying that software is a small part of this work
The financial concerns will be just the tip of the iceberg

Learning Objectives

- Identify challenges to the successful rollout of enterprise systems in organizations
- Describe different models for the rollout of enterprise systems
- Discuss the pros and cons of different rollout options
- Elaborate on important prerequisites to the successful rollout of enterprise systems

Characters

Regina Wood—VP, Eastern MA
Anik Malik—VP, New Hampshire
Samantha Bellman—VP, Western MA
David Ashley—VP, Rhode Island
Kevin Sanders—Scheduling/Dispatch, New Hampshire
Jamie Cochrine—Operations, Rhode Island
Claudia Narez—Environmental, Eastern MA
Abra McGregor—Finance, Western MA

Narrator: *In the last meeting, part of our group considered the possibility that processes may be bought as part of an enterprise system and modified. That meeting ended with the group expressing some concerns. Regardless, the department heads have been continuing the dialogue with the VPs. They've realized it will take considerable effort to roll out whatever they buy—and modify. They're not sure if it's simply a matter of how easy the software is, whether it is about how the employees accept or reject the software, or something else. This is what they want to explore today. In this meeting, we have the VPs of all four divisions as well as those department heads who've been participating in this effort all these months. The group is aware that they may, in the end, still end up redesigning only a few key processes, and for that, they may get help from* *the Sonita BPMS. They are becoming increasingly comfortable with the idea of acquiring an enterprise system like TBQ, using their prebuilt processes, and modifying them a little for Royal Energy. But they're worried about the scale of the effort. This is a critical meeting. The group wants to know what challenges they may face when they try to roll out the processes as part of an enterprise system like TBQ.*

ANIK: Alright, I want to be the first one to say it. I am so glad we're here.

SAMANTHA: I second that!

(general sounds of agreement)

ANIK: I mean, we're finally getting a solution to this problem we've been struggling with forever. It feels great.

SAMANTHA: Anik, you're positively radiant.

DAVID: Well, don't count your eggs before we know what hens we want, or whatever . . . we still have to figure some stuff out.

ABRA: We know there will be some challenges as we get to roll out—we need to know what they can be so we are prepared.

REGINA: That's true. Okay—Abra, Claudia, Kevin, Jamie. What do you have for us? We're listening.

CLAUDIA: Okay. Well, we've done some research. We've looked at the stats from other companies who chose to implement TBQ—or for that matter, any other enterprise system—and as a part of that, roll out new business processes. And of course, we consulted Gartner research.

ANIK: Due diligence!

CLAUDIA: Exactly. We left no stone unturned. But I'm afraid the results have left us . . . well, somewhat ambivalent about this whole thing.

SAMANTHA: Well, I for one am definitely curious about what you've found, but let's keep it real here. Now that acquiring TBQ is turning out to be a realistic alternative, we need to be aware of any and all mistakes other companies have made. That'll help us be better prepared.

ANIK: True. So don't mince words. Tell us exactly how it is.

CLAUDIA: Fortunately, that's what we've planned. But we have a more balanced presentation for you today.

ABRA: Kevin is going to bring up the challenges and concerns, and we will have Jamie talk about the positives.

KEVIN: I love being the bad guy.

DAVID: Well, let's have it! Do your worst.

KEVIN:	Okay, so . . . let me start with this bombshell: the track record of implementing enterprise systems is definitely not good. There are a lot of large failures. It also costs a lot—often multiple times the original budgets.
REGINA:	For example . . . ?
KEVIN:	Well, instead of the planned two million, it could end up being thirty million.
REGINA:	Dang, Gina!
KEVIN:	Yeah. And instead of the planned eighteen months, it could be six years. And even when the software is implemented, it can sometimes remain hard to use.
SAMANTHA:	Ouch.
DAVID:	I know. I can't believe Regina made a *Martin* reference.
REGINA:	You *can't?*
DAVID:	Anyway, this news reminds me of one of the early failures that's been written up so many times. Remember what the Hershey Company went through?
ANIK:	Oh, the Great Chocolate Debacle.
DAVID:	Yes! They spent years trying to roll out new software. Employees balked, operations were derailed, customers suffered, their stock price tanked—it was total chaos.
ANIK:	Right, I've heard of a few similar cases with other companies.
JAMIE:	So yes, Hershey is the case everyone remembers, but that was a long time ago. Software developers and implementation consultants have learned a lot since then. There is some wisdom in the industry about what needs to be done to improve our chances of success.
REGINA:	Like what?
JAMIE:	Well, we've learned that we have to include the stakeholders and users right from the get-go. We can't just let IT do it by themselves. Yes, we know that an enterprise system like TBQ is a complicated piece of software . . . but it's not enough to implement it. It is even more important to make sure we include the current and future users in the planning stages.
REGINA:	Like when we are deciding the implementation schedule?
JAMIE:	Well—when we are making the decisions about how to modify processes, what can and cannot be done with the prebuilt processes! So perhaps I should say the design stage, not the planning stage?

	Well, in any case, before it's all done and dusted and just before we move to rollout.
ANIK:	That makes sense. So when should we start including users? And we include managers as well, right?
JAMIE:	Yes! Right from the beginning. Not later or as an afterthought; right at the get-go.
SAMANTHA:	Will they understand what the software is, what it can do, what we're trying to do? Should we not wait till a little later and train them after the software is installed?
JAMIE:	That's exactly it! We can't wait that late in the cycle. They must participate in how the software will be configured and customized. And that means they must have some say in whether we will be able to use those built-in processes as they are within TBQ or, if they need to be modified, how they should be modified.
SAMANTHA:	Hmm . . . but—
JAMIE:	Well, look at it this way. If we don't have our users as part of these design stages, they're only going to see the processes after we accept the built-in version or modify. And then they don't know what the issues are.
SAMANTHA:	Like this problem of new software versions?
JAMIE:	Exactly! So this is what the research is showing. We need to bring the users in during those early stages and have them become part of those design decisions, so if we end up with prebuilt processes that are different from the current processes, they know *why* we have to move to a new process.
SAMANTHA:	That's a sound argument. And if they're so familiar from the get-go, that'll save on training time.
JAMIE:	Oh no, we will definitely still do training—we have to. And a lot of it. This training must be completely thorough *and* targeted to different users. It should be available to users multiple times, not just once.
SAMANTHA:	What do you mean by that?
JAMIE:	Well, from what I understand, enterprise systems like TBQ tend to be quite complex. So, for users, there is much on screen that they need to navigate. And they're not going to get a good sense of what they are doing unless it's shown in a step-by-step manner, many times. And more importantly, they have to understand why their actions are important to someone else doing the next task in the process someplace else in the company.

SAMANTHA: (*nodding*) So we explain to them *why* they're working with the TBQ interface, not just how to do their own job! I understand—what you're saying is that we can talk to the users about what can and *cannot* be modified, both the interface and the process.

ABRA: (*sensing that we need to move to another point*) The other important thing is communication and assurance. We have to clearly tell our current employees—the future users of TBQ—that they are *not* going to lose their jobs because of the TBQ rollout.

DAVID: We have good people; we want to keep them!

ABRA: Correct. So we want to clearly let them know that even if we go with some of the prebuilt processes in TBQ and those are different from our current work practice, that's not going to lead to anyone getting fired. We will help and train them to get good at using TBQ.

REGINA: Wow . . . yes. I can see how this can be a concern.

JAMIE: It's very important that we continue to communicate this assurance throughout the project. Let me add a phrase I picked up in my research. What we need to do is emphasize that we want our employees to be *co-owners* of this change we're pursuing.

REGINA: I like that term, *co-owners*.

JAMIE: So we have to have a different mindset. This is *not* a matter of installing some new software! This is about bringing in a large change at Royal Energy. And we have to treat our work accordingly—for us, this is change management!

REGINA: Good point, Jamie!

JAMIE: So as a part of our effort, our employees have to be confident that top management is completely supportive of this change and that we have allocated sufficient resources to make it happen.

SAMANTHA: Wow . . . I like what I'm hearing. Good work, people.

REGINA: I agree. I love the emphasis on communication. It's our people who'll end up actually using the software after it's rolled out. They're the ones who need to be comfortable with it.

DAVID: Right, these are the details we should be focusing on. I mean, our users are spread out across different places. Everyone in each department has to be able to do tasks in basically the same way with this new software. But we can't just push it down on them. We want to invite them to be part of that design effort. Otherwise, I can see how we'll repeat those mistakes that Hershey made! That's the whole point of our research, isn't it?

ANIK: So . . . my concern is that we have some people in the company who are older and have been with us for years. How do we get them to buy into what we're trying to do here? Remember, we have a lot of special processes . . . can we meet our people where they are and make the effort to configure and customize TBQ's prebuilt processes without frustrating them?

REGINA: Exactly, what should we do to make sure that our decisions for modifying processes are appropriate for the people we have?

ABRA: We need to remember that we aren't really buying TBQ as just a piece of software. It's more like we're buying their way of doing business.

(collection of "Oh" and "Ah, I see")

SAMANTHA: I remember this point from Sophie—last meeting, was it? And Jamie also said it in a different way earlier. It's not that we are just buying their software; it's almost like we're buying *into* their way of doing business.

DAVID: But we're unique in so many ways. That's what makes us special and different. We've developed a loyal customer base by doing things a certain way. Will we lose that?

REGINA: But remember how much we've been suffering lately. That's why we started this whole effort! Maybe what we did worked well back then, but now we need to be more open to different ways of doing things.

DAVID: Okay, I'll grant you that. But if all companies buy the same software from TBQ, what makes us different? Are we doing the same things that our competitors may be doing? How does that help us?

(pause . . . then . . .)

CLAUDIA: I have to think that it's our people. We've worked to establish trust and confidence in our workforce. We've recruited well, and some of our employees treat Royal Energy as their life's work. None of these processes are going to deliver value to our customers unless we have talented, committed, and caring employees. We have those people on our team.

JAMIE: Yes, so what we have to do is to ensure that TBQ does not stand in the way of doing our work—it speeds it up, makes it more efficient! We look at TBQ as an opportunity to strengthen qualities we've always had.

DAVID: It's almost like this software is a love letter to the company . . .

(silence)

DAVID: Is that maybe taking it too far?

REGINA: (*chuckles*) Well, we're investing in the potential of our company. And our company is nothing without the people. We are essentially buying into how TBQ thinks a utility company like ours should operate. We are already great; our people make us great. And now we're giving them the boost they need to be so much more. To make this work, it's going to take a lot of effort across the board. But I believe it will be worth it.

ANIK: Well, all of this is a good discussion so far, but I want to acknowledge that we are talking about a big change in our company. How do we really do this? One fine day, do we just say "Out with the old and in with the new"?

SAMANTHA: I've been playing that in my head as well. Anik, Jamie, Kevin—did you find anything about this in your research?

JAMIE: I remember this part as we did our research. (*consulting his notes*) Yes, here you go! There are two main approaches to rolling out TBQ and the set of processes. One is the big bang approach, and the other is the phased approach.

SAMANTHA: Tell us more, please!

JAMIE: Well, the big bang approach is just that. We prepare well, we train everyone, and then, one fine day, we decide to just switch from the old to the new—for the entire company.

DAVID: Oh boy! But what if things go wrong?

ABRA: So that is exactly why we *must* plan and get ready.

DAVID: You're right! We have to make sure that the software mods are tested, the rollout plans are firm, and we have everyone trained. And we have to make sure we have migrated all the data from the old systems to TBQ. So when we move to TBQ, people would come in on, say, June 1 and they would start using TBQ and the new processes. Why not do it in small increments?

ABRA: Well, the advantage is that we're not maintaining two distinct sets of data and systems.

KEVIN: Now *that* would be a nightmare, wouldn't it?

SAMANTHA: (*seeing the wisdom in this option*) Yes, I see that now!

DAVID: Humor me! Could we do small increments?

JAMIE: Yes, we could. Think of it this way: We can either have some of the departments go to TBQ first, try it out, and then bring on the other departments. Or we can have some of the processes follow the prebuilt templates in TBQ and then bring along other processes. Either way, we are going to have individuals and groups dealing with multiple systems simultaneously.

SAMANTHA: I see! So do companies ever use this sort of incremental approach?

JAMIE: Phased approach is what they call it, and yes, some companies may use it, particularly when they have some processes that are so critical that human lives depend on them, like in a hospital or other critical industries. They may even run the old versions of the process and the new ones from TBQ simultaneously until they feel confident. But as more companies have implemented enterprise systems, this approach is increasingly rare. Many companies seem to favor the big bang approach with a cutover from the old to the new.

SAMANTHA: Well, all of this seems to be more difficult than we expected. We'll need to plan well and train everyone, but I'm becoming convinced. Is there anything we need to think of to ensure that things don't go too wrong when we cut over from the old to the new?

JAMIE: Yes there are! Here's one. One of the successful cases we read about—they had a very clever but simple approach. They essentially retained the services of the consultants who were responsible for the software mods and process customization well past the cutover date. So the consultants were asked to be "on call" for six months past the rollout date. That, along with a sustained campaign of early user involvement and training, meant that the company planned the actual rollout very well and was well prepared in case something went wrong.

SAMANTHA: Love that!

DAVID: Agree! We need this kind of insurance!

ANIK: It looks like bringing in TBQ will be risky, but we have a good sense of what may go wrong, and the positives far outweigh the risks. I'm in!

DAVID: You can count me in.

KEVIN: A coffee toast! To the people!

ALL: To the people!

The End

Reflection Questions

1. When the scenario begins, Kevin starts with a bombshell: "The track record of implementing enterprise systems is definitely not good. There are a lot of large failures. It also costs a lot—often multiple times the original budgets." As the discussion progresses, the group realizes that their mindset needs to change. It is not merely "software implementation." What is the mindset Royal Energy needs to adopt? Explain.

2. To improve the probability of success, the group discusses the importance of communication. Abra introduces this in the conversation: "The other important thing is communication and assurance. We have to clearly tell our current employees—the future users of TBQ—that they are *not* going to lose their jobs because of the TBQ rollout." What should be the elements of a communication strategy at Royal Energy? For example, these elements may include (a) providing information about upcoming changes, (b) providing assurances that the changes will not lead to job losses, and what else? The group seems to have touched upon half a dozen of these through their conversation. Bring these together in your answer and explain.

3. It is Anik who brings this up later in the conversation: "One fine day, do we just say 'Out with the old and in with the new'?" In response, Jamie suggests two options for Royal Energy to consider for rolling out TBQ and the new processes. It is useful to think of those as two anchors, so a company like Royal Energy can craft their own approach somewhere on this continuum. What criteria would you use to craft this approach for Royal Energy? How would your decision be different for your own company?

Self-Assessment Questions

1. Can you identify challenges to the successful rollout of enterprise systems?
2. Can you describe and assess different approaches to the rollout of enterprise systems?
3. Can you elaborate prerequisites to the successful rollout of enterprise systems?

Readings

1. Ali, M., and L. Miller. 2017. "ERP System Implementation in Large Enterprises: A Systematic Literature Review." *Journal of Enterprise Information Management* 30 (4): 666–92.

2. Huq, Z., F. Huq, and K. Cutright. 2006. "BPR through ERP: Avoiding Change Management Pitfalls." *Journal of Change Management* 6 (1): 67–85.

Key Terms

Big Bang Implementation

Big bang implementation describes the situation when a new *enterprise system* is launched all at once, at a point in time. All users switch from the old to the new system, and all functionalities go live. The transition requires extensive preparation, testing, and training before the launch.

Data Conversion

Data conversion refers to the process of transferring existing data from the old system(s) to the *enterprise system*. It involves several tasks such as data identification, data extraction, data transformation, data loading, and data verification. These can be challenging because of issues such as compatibility and the large volume of data that an organization possesses.

Phased Implementation

Phased implementation describes the situation when the rollout of a new *enterprise system* is broken down into phases. Instead of a big bang, the rollout is planned over time—for example, some modules at a time, some locations at a time, or some departments at a time. This approach may be less risky but also more complex because the old and new systems run concurrently for some time, which may cause inconsistencies.

Afterword
Reflections

Thank you, good folks, for sharing this composition
We hope you feel somewhat prepared for your own expedition

We have been active listeners throughout this journey that Royal Energy has taken. The ten scenes across three acts have allowed us to witness how different players within Royal Energy have participated in raising issues and making decisions. As we have moved from one scene to the next, we have also been privy to some of the documents and working papers they have generated. Clearly, we do not have the complete set. That would have been too overwhelming for us. But the examples we have seen tell us how different players at Royal Energy have participated in the journey. We hope that you have been following up on the questions at the end of each scene to reflect on how Royal Energy has responded to the challenges they have faced. Now Royal Energy is moving ahead with its rollout plans.

We leave you now to reflect on what you have learned and how to move to the next adventure. We hope that the self-assessment questions, key terms, and readings for each scene will help you determine if you were able to pick up important lessons and revisit them as needed. Happy hunting!

Index

Page numbers in *italics* and **bold** refer to figures/tables and glossary terms, respectively.

www.ingramcontent.com/pod-product-compliance
Lightning Source LLC
Chambersburg PA
CBHW061324220326
41599CB00026B/5025